AF581857

Climate and Atmospheric Dynamics and Predictability

Climate and Atmospheric Dynamics and Predictability

Editors

Ioannis Pytharoulis
Petros Katsafados

MDPI • Basel • Beijing • Wuhan • Barcelona • Belgrade • Manchester • Tokyo • Cluj • Tianjin

Editors
Ioannis Pytharoulis
Aristotle University of Thessaloniki
Greece

Petros Katsafados
Harokopio University of Athens
Greece

Editorial Office
MDPI
St. Alban-Anlage 66
4052 Basel, Switzerland

This is a reprint of articles from the Special Issue published online in the open access journal *Climate* (ISSN 2225-1154) (available at: https://www.mdpi.com/journal/climate/special_issues/Atmospheric_Dynamics).

For citation purposes, cite each article independently as indicated on the article page online and as indicated below:

LastName, A.A.; LastName, B.B.; LastName, C.C. Article Title. *Journal Name* **Year**, *Volume Number*, Page Range.

ISBN 978-3-0365-0224-3 (Hbk)
ISBN 978-3-0365-0225-0 (PDF)

Contents

About the Editors

Ioannis Pytharoulis is Associate Professor in the School of Geology of the Aristotle University of Thessaloniki (AUTH; https://www.auth.gr/en), Greece and the director of the Laboratory of Meteorology and Climatology of AUTH (LMC-AUTH). His main research interests lie in numerical weather prediction and synoptic/mesoscale meteorology, focusing on intense weather events. He is responsible for the operational weather forecasting system of LMC-AUTH (http://meteo.geo.auth.gr). He received a Bachelor's degree in Mathematics from AUTH, and he received an M.Sc. in Weather, Climate and Modelling and a Ph.D. in Meteorology from the Department of Meteorology of the University of Reading, United Kingdom. He was formerly employed as a postdoc researcher in the Atmospheric Modelling and Weather Forecasting Group of the Physics Department of the University of Athens in Greece, a meteorologist in the Numerical Weather Prediction Division of the Hellenic National Meteorological Service, a laboratory teacher at the Technological Educational Institute of Piraeus in Greece, a weather forecaster and scientific consultant in the private sector and Lecturer & Assistant Professor at LMC-AUTH. He has participated (as PI or researcher) in 30 research projects funded by various national and international authorities. He has been a Guest Editor for three Special Issues of the international journals *Atmosphere* and *Climate*, as well as an Associate Editor of the international journal *Acta Geophysica*. A more detailed CV appears at https://users.auth.gr/pyth/.

Petros Katsafados has been Associate Professor at the Department of Geography of Harokopio University of Athens since 2007 and is head of the Atmosphere and Climate Dynamics Group (ACDG; http://meteoclima.gr). He originally studied Mathematics but then he switched scientific discipline to Atmospheric Physics and Dynamics and he finally completed his master's degree at the School of Physics at the National and Kapodistrian University of Athens (NKUA). His Ph.D. thesis focused on factors that influence the predictability of numerical weather prediction. Currently, he is author or co-author of more than 40 articles in peer-reviewed scientific journals and more than 85 peer-reviewed conference papers. His work has received more than 1000 citations (h-index = 17; source: Scopus). He is also author of two textbooks, and he has contributed to several book chapters and special editions. A more detailed CV is available at https://www.geo.hua.gr/wp-content/uploads/Katsafados-Petros-cv-en.pdf.

Preface to "Climate and Atmospheric Dynamics and Predictability"

The state of the weather and climate is largely defined by the interactions between the various components of the climate system (atmosphere, hydrosphere, land surface, cryosphere, biosphere). The understanding and the prediction/projection of the atmospheric and climate dynamics (i.e., how the natural laws determine the weather and climate) are essential for life, property and environment. High-impact weather systems, low-frequency oscillations and their climatic variability exert a significant influence on humans and their activities. Over the last few decades, the advances in weather and climate numerical models and the increase of computational resources have resulted in a blooming of weather forecasting and climate research, allowing for more effective planning and preparedness against adverse weather and climate change.

This Special Issue was mainly devoted to collecting observational, theoretical and modeling studies on the dynamics of the atmosphere and the climate system, as well as on their predictability at different spatiotemporal scales.

Ioannis Pytharoulis, Petros Katsafados
Editors

Article

Development of a Front Identification Scheme for Compiling a Cold Front Climatology of the Mediterranean

E. Bitsa [1,*], H. Flocas [1], J. Kouroutzoglou [2], M. Hatzaki [3], I. Rudeva [4] and I. Simmonds [4]

[1] Department of Physics, National and Kapodistrian University of Athens, 15784 Athens, Greece
[2] Hellenic National Meteorological Service, 16777 Athens, Greece
[3] Department of Geology and Geoenvironment, National and Kapodistrian University of Athens, 15784 Athens, Greece
[4] School of Earth Sciences, University of Melbourne, Parkville, Victoria 3010, Australia
* Correspondence: ebitsa@phys.uoa.gr

Received: 30 September 2019; Accepted: 6 November 2019; Published: 11 November 2019

Abstract: The objective of this work is the development of an automated and objective identification scheme of cold fronts in order to produce a comprehensive climatology of Mediterranean cold fronts. The scheme is a modified version of The University of Melbourne Frontal Tracking Scheme (FTS), to take into account the particular characteristics of the Mediterranean fronts. We refer to this new scheme as MedFTS. Sensitivity tests were performed with a number of cold fronts in the Mediterranean using different threshold values of wind-related criteria in order to identify the optimum scheme configuration. This configuration was then applied to a 10-year period, and its skill was assessed against synoptic surface charts using statistic metrics. It was found that the scheme performs well with the dynamic criteria employed and can be successfully applied to cold front identification in the Mediterranean.

Keywords: cold fronts; climatology; Mediterranean; identification scheme; Frontal Tracking Scheme (FTS); MedFTS

1. Introduction

Cold fronts are significant components of the weather and climate systems, and can be closely associated with extreme events. The passage of a cold front is indicated by, and associated with, substantial variations of temperature, humidity and wind. The identification of cold fronts has attracted more scientific interest than their warm counterparts because of their more discrete character and their connection with severe weather phenomena [1–6].

Despite the availability of numerical prediction and analysis models, the manual identification of fronts on weather charts is a time-consuming process that introduces a high degree of subjectivity, even for an experienced operational meteorologist [7,8]. The complexity of the task dictates that the compilation of frontal climatologies by manual methods is not feasible. Hence, there is practical and scientific interest in developing automated schemes to create such climatologies from observed data, reanalyses, and climate model outputs [8]. The advantage of automated detection methods is that they are objective, reproducible, and fast.

The majority of such objective and automated front identification methods in the literature use thermal criteria [6,9–13]. Most of these studies focus on large-scale fronts that develop and move across vast areas of oceans and continents. Thus, their identification is facilitated by the large scale of fronts and the homogeneity of surfaces. However, verification of these algorithms has shown that the recognition of the frontal surfaces, taking into account only temperature gradients, is inadequate in many cases for complex features [14,15]. Some of these methods are used routinely in weather forecasting [16], whereas others focus on extreme events, such as widespread fires [17]. A number of automated algorithms have been applied in order to generate frontal climatologies for southwest Western Australia [18], for the globe [7], and for the Southern Hemisphere [19,20].

Since the Mediterranean Sea is a closed basin surrounded by complex topography, its fronts tend to exhibit small spatial and temporal scales, as well as complicated kinematic and thermodynamic features during their lifetime [21]. Climatological studies focusing on the Mediterranean fronts are relatively few, and the early studies were based on subjective approaches utilising synoptic surface maps [22]. The investigation of a nine-year period (1971–1979) of daily charts [22], demonstrated that fronts appear very frequently in the Mediterranean throughout the year with maximum frequency one every seven days in winter. While identification schemes have been applied to diagnose the climatologies of cyclonic [23–25] and anticyclonic centers [26] in the Mediterranean, there is no corresponding application for the analysis of cold fronts.

The objective of this study was to develop and evaluate a scheme for the identification of cold frontal systems in the Mediterranean basin which is based on the Frontal Tracking Scheme (FTS) [19]. In Section 2, the scheme and the modifications performed are presented in brief, while in Section 3, typical results of the sensitivity tests are given for specific high impact cases connected with cold fronts passages over the Mediterranean. In Section 4, a statistical validation of the scheme for a decade is given and, finally, in Section 5, the main conclusions are summarized.

2. Description of the Identification Scheme

FTS was developed at The University of Melbourne, Australia [19], and has been used for the climatological study of Southern Hemisphere cold fronts. Unlike other similar schemes which use thermal criteria [7,10,16], FTS uses only wind-related criteria to identify fronts and has proved to work well compared to the other schemes. More specifically, thermal based methods are known to have difficulties identifying fronts in the areas of high temperature contrasts, such as coastal areas and regions with elevated topography [15]. Hence, the Mediterranean region would be a particularly difficult site for frontal identification using thermal variables. Furthermore, thermal based methods may not be able to reliably distinguish between cold/warm fronts [27].

FTS is based on Eulerian changes of the 10 m meridional wind component (v) which is valuable in diagnosing various aspects of frontal behavior [8,28]. The criteria for identification are [19]: (a) at a time t, grid points are flagged where the wind changes from the southwestern quadrant (westerly zonal wind $u > 0$, southerly meridional wind $v > 0$) to the northwestern quadrant (westerly zonal wind $u > 0$, northerly meridional wind $v < 0$) between subsequent time points t and (t + 6 h), (b) the change of the meridional wind dv exceeds a specific threshold value dv_{crit} during the same 6 h period.

The grid points which satisfy the above-mentioned criteria are flagged and a component labelling technique is applied [29]. Then, each flagged pixel is related and connected to its nearest eight neighbors, giving clusters of grid points. The location of the front is determined by the eastern edge of each cluster. As this approach is applied to all of the eastward edge points, the output is a set of latitude and longitude points which mark the location of the front. Thus, the location of the front at the time t + 6 h is given by a single series of longitude values. These values will have a stepwise character, since they represent discrete grid points. For this reason, the longitude values are treated as a simple series and smoothed by a resistant smooth method [30] appropriate for equally-spaced data. This robust statistical technique employs a set of short-window running median and running mean filters, which are successively applied multiple times to the series, to achieve adequate smoothing. Then, the obtained smoothed eastern edge determines a "mobile front". This method is particularly suited for the detection of strongly elongated, meridionally oriented fronts, which typically extend far from a cyclone center [19].

Since FTS was developed to identify cold fronts in the oceans while the topography of the Mediterranean affects the formation and characteristics of cold fronts [31], in this study FTS is modified to better identify the position, scale and tilt of cold fronts in the Mediterranean. From the records of the Hellenic National Meteorological Service, twenty cases of cold fronts are selected that entered Mediterranean from different regions (e.g., Atlantic, North Africa, northern Europe) or formed in different parts of the Mediterranean (Western, Central and Eastern) during different months throughout the year, having caused intense precipitation.

The initial criterion used in MedFTS is that the zonal component u is westerly both at t and t + 6 h and the meridional wind changes sign from positive to negative. Then, sensitivity tests were performed on the other wind related criteria. First, sensitivity tests are made on the criterion of meridional wind change within the 6 h time step (dv), in order to find the optimum threshold value of the meridional wind magnitude change dv_{crit}. Second, instead of using the change of the meridional wind component (dv), the shift of the vector wind direction is employed during a 6 h period ($d\varphi$), where φ = arctan (v/u) and a specific minimum threshold value $d\varphi_{crit}$ is also investigated. This criterion is examined to better identify zonally elongated cold fronts and at the same time, to filter out erroneously identified front segments. Third, an additional criterion of the magnitude of vector wind $|U|$ exceeding a specific critical threshold $|U|_{crit}$ in each cluster of grid points is added to optimise the scheme, considering the operational experience of forecasters that the intensity of the northwesterly wind is significant behind the cold front. This criterion allows the discarding of shallow fronts.

3. Sensitivity Tests

For the twenty selected cases, ECMWF Re-Analysis (ERA)-Interim datasets of zonal u and meridional v wind components at the near-surface level (10 m) on a 0.5° × 0.5° resolution are used [32]. This high resolution was chosen in order to obtain a better representation of the small-scale fronts appearing in the Mediterranean region. The scheme results are compared with the surface analyses produced by the UK MetOffice, archived by www2.wetter3.de, available every 6 h. Furthermore, for the validation of the results, MSG IR 12.0 μm satellite images, available from the Hellenic National Meteorological Service, are employed. We restrict ourselves here to showing results for the case of 7–10 November 2016 that included two extended tilted cold fronts that travelled across Mediterranean. However, the results were consistently investigated and validated for the other cases. The tests are performed following the rationale described in Section 2.

First, the critical values of the 6 h meridional wind change dv_{crit} was explored. An initial threshold value dv_{crit} = 2 m s^{-1} was employed, as suggested by [19] in the initial version of FTS. Then, the scheme was employed for different values of dv_{crit} increasing by 1 m s^{-1}. Figure 1a shows the synoptic situation of 7 November 2016, 00:00 UTC. In Figure 1b, the identified fronts are depicted (red lines) for the same day and hour for dv_{crit} = 3 m s^{-1}. In the same Figure, the light blue regions show the areas where the wind shift criterion is satisfied.

A comparison of Figure 1a,b reveals that the scheme succeeds in identifying fronts over the Atlantic. However, in the Mediterranean, its performance is somewhat lower: although it identifies correctly the main cold front over Italy, this is segmented over the Adriatic Sea while other frontal fragments are produced which do not exist in the synoptic analysis. For larger values of dv_{crit} of 4 and 5 m s^{-1}, the erroneous front identifications show a tendency to diminish (Figure 1c). However, when dv_{crit} exceeds the value of 6 m s^{-1}, the existing fronts incline to be broken into smaller fragments (Figure 1d). Therefore, moderate values of dv_{crit} ranging between 4–6 m s^{-1} seem to best represent Mediterranean cold fronts. Similar results were derived for the following hours. Figure 2 shows the results for 8 November 2016 at 12:00 UTC.

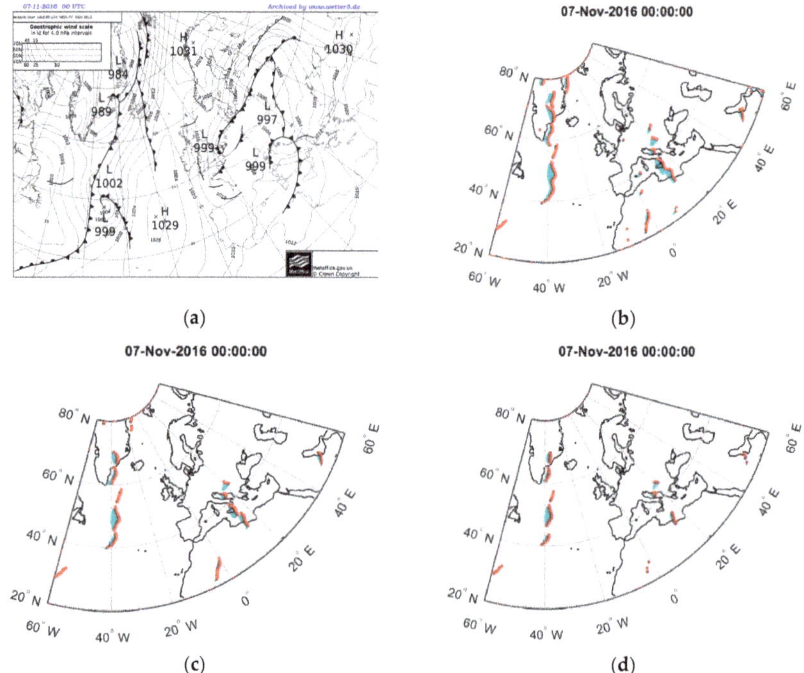

Figure 1. (**a**) Synoptic surface chart over the area of interest at 00:00 UTC 7 November 2016, and identified fronts for (**b**) dv_{crit} = 3 m s^{-1}, (**c**) dv_{crit} = 5 m s^{-1}, and (**d**) dv_{crit} = 7 m s^{-1}. The red lines show the identified fronts, whereas the areas where the wind shift criterion is satisfied are depicted with the light blue color.

Since the analysed Mediterranean fronts are not purely meridionally elongated, but they rather tend to assume a more zonal orientation, the 6 h change of the total wind direction $d\varphi$ is explored instead of the change of the meridional wind magnitude. It should be noted that this criterion was found helpful in identifying cold fronts that present zonal orientation in the beginning of their life [31] and the criterion on dv might not be able to properly identify them. The scheme was tested for different values of $d\varphi_{crit}$ starting from 20°, with steps of 10°. In Figure 3, the identified fronts are depicted for 7 November 2016, 00:00 UTC, for (a) $d\varphi_{crit}$ = 30° and (b) $d\varphi_{crit}$ = 50°. A comparison with the synoptic analysis in Figure 1a shows that the scheme indeed represents the front over Italy when $d\varphi_{crit}$ = 30°. However, it can be appreciated that using solely the $d\varphi$ criterion, several erroneously identified fronts are obtained. Moreover, for $d\varphi_{crit}$ = 50°, the front over Italy is mistakenly broken into smaller fragments. Similar results were obtained for $d\varphi_{crit}$ > 40° and the best representation of fronts was obtained for $d\varphi_{crit}$ = 30° in most cases. To effectively filter out the erroneously identified frontal objects without losing the spatial continuity of the correctly identified fronts, the value $d\varphi_{crit}$ = 30° is adopted.

The use of the $d\varphi$ criterion alone is not adequate, since it may lead to mistakenly identified fronts in cases when, at t + 6 h, the wind is weak or there is calm. For this reason, another criterion that we apply is that the maximum magnitude of the vector wind is above a threshold value $|U|_{crit}$. It should be noted that apart from filtering out mistaken identifications, this criterion also helped in identifying cold fronts entering from North Africa and distorted cold fronts in the Central Mediterranean that rejuvenated when entering the Aegean Sea. To find the optimum value of $|U|_{crit}$, values between 0 and 10 m s^{-1} were tested. Figure 4 depicts the results of the scheme on 7 November 2016 00:00 UTC for (a) $|U|_{crit}$ = 5 m s^{-1} and (b) $|U|_{crit}$ = 7 m s^{-1}. Red lines represent the identified fronts, whereas colored areas show the magnitude of the total wind $|U|$ at the grid points where the $d\varphi$ criterion is met. It is clear that a value of $|U|_{crit}$ = 5 m s^{-1} effectively filters out spurious front identifications, while a larger value (e.g., $|U|_{crit}$ = 7 m s^{-1}) erroneously filters out the front over Italy. From the above sensitivity tests, the combination of $d\varphi_{crit}$ = 30° and $|U|_{crit}$ = 5 m s^{-1} seems to best represent cold fronts in the Mediterranean at each following synoptic time (Figures 5 and 6). Similar results were derived for the other selected cases under a variety of synoptic environments. It should be noted that based on these dynamic criteria, warm frontal structures are not identified.

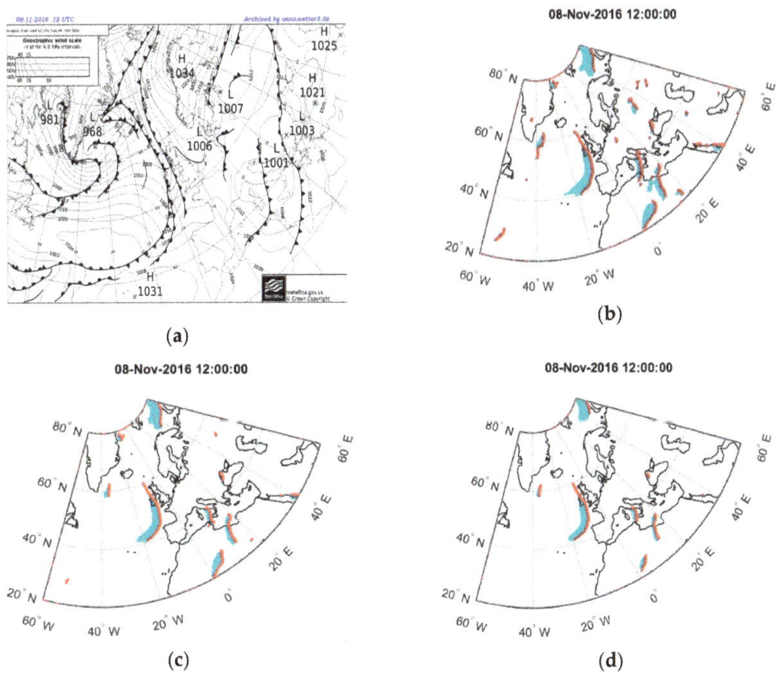

Figure 2. (**a**) Synoptic surface chart over the area of interest at 12:00 UTC 8 November 2016, and identified fronts for (**b**) dv_{crit} = 2 m s^{-1}, (**c**) dv_{crit} = 4 m s^{-1}, and (**d**) dv_{crit} = 6 m s^{-1}. The red lines show the identified fronts, whereas the areas where the wind shift criterion is satisfied are depicted with the light blue color.

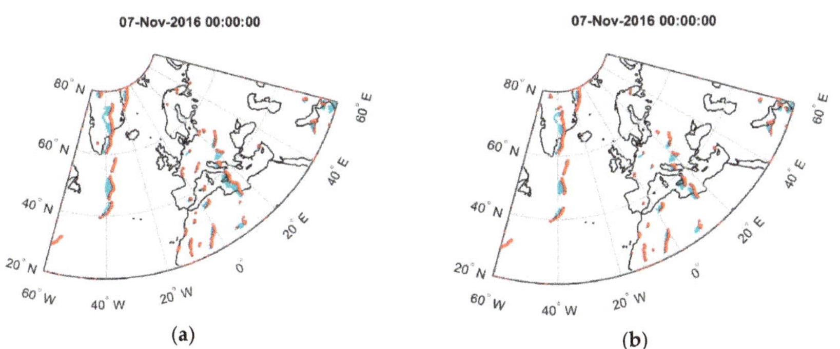

Figure 3. Identified fronts at 00:00 UTC 07 November 2016, for (**a**) $d\varphi_{crit}$ = 30° and (**b**) $d\varphi_{crit}$ = 50°. The red lines show the identified fronts, whereas the areas where the wind shift criterion is satisfied are depicted with the light blue color.

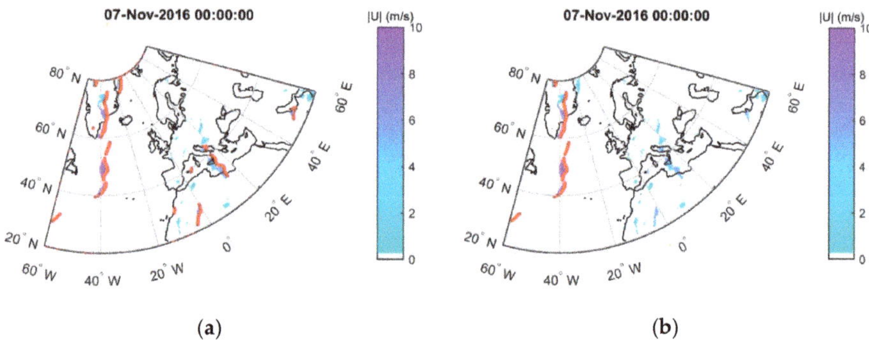

Figure 4. Identified fronts at 00:00 UTC 7 November 2016, for $d\varphi_{crit} = 30°$, (**a**) $|U|_{crit} = 5$ m s^{-1} and (**b**) $|U|_{crit} = 7$ m s^{-1}. Red lines represent the identified fronts, whereas colored areas show the magnitude of the total wind $|U|$ at the grid points where the $d\varphi$ criterion is met.

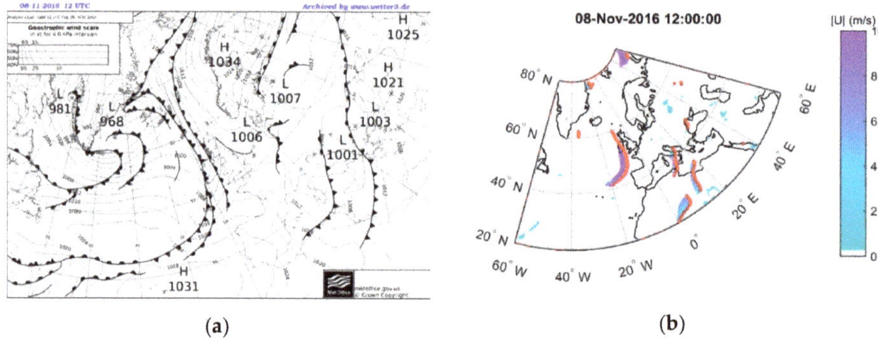

Figure 5. (**a**) Synoptic surface chart at 12:00 UTC 08 November 2016, and (**b**) identified fronts for $d\varphi_{crit} = 30°$ and $|U|_{crit} = 5$ m s^{-1}. Red lines represent the identified fronts, whereas colored areas show the magnitude of the total wind $|U|$ at the grid points where the $d\varphi$ criterion is met.

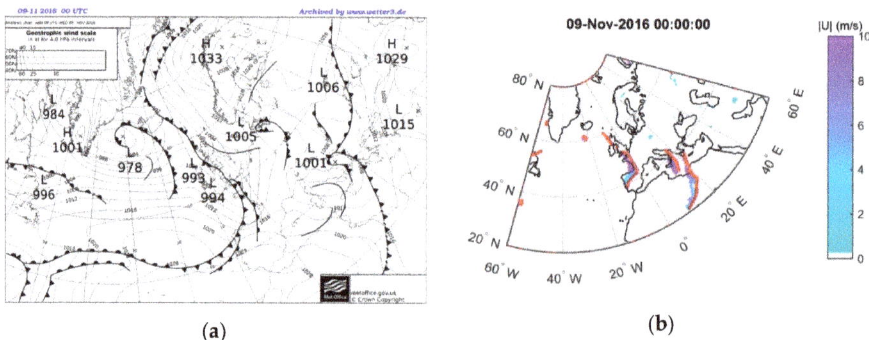

Figure 6. (**a**) Synoptic surface chart over the area of interest at 00:00 UTC 9 November 2016, and (**b**) identified fronts for $d\varphi_{crit} = 30°$ and $|U|_{crit} = 5$ m s^{-1}. Red lines represent the identified fronts, whereas colored areas show the magnitude of the total wind $|U|$ at the grid points where the $d\varphi$ criterion is met.

In summary, in order to identify a front in our MedFTS scheme we require: (a) the zonal component u is westerly both at t and $t + 6$ h, (b) the meridional wind changes sign from positive to negative, (c) the directional shift of the wind ($d\varphi$) exceeds the threshold of $d\varphi_{crit} = 30°$ and (d) the magnitude of the vector wind $|U|$ is greater than $|U|_{crit} = 5$ m s^{-1}. It should be noted that the initial criterion of $dv_{>} dv_{crit}$ used in FTS has been replaced by both the criteria of $d\varphi > d\varphi_{crit}$ and $|U| > |U|_{crit}$.

In order to check the performance of MedFTS, the results for 19 March 2018 at 12 UTC is demonstrated in Figure 7 which included a cold front that developed over Tunisia and affected Greece after rejuvenation and a zonally oriented cold front over the Iberian Peninsula that entered from the Atlantic and moved towards western Mediterranean. Figure 7a,b shows that the surface analysis agrees well with the satellite image for the cold front over Greece. The scheme correctly represents the location and orientation of the fronts at this specific time (Figure 7c), avoiding many erroneous identifications before the use of the $|U|_{crit}$. The zonally elongated front over the Iberian Peninsula is also identified correctly by the scheme (Figure 7c) along with its subsequent evolvement. Figure 7d presents the results when the dv criterion is solely used. From the comparison between Figure 7c,d becomes evident that the zonal front over the Iberian is not properly identified with the dv criterion. Therefore, it is suggested that the successful identification of this front is mainly attributable to the combination of $d\varphi_{crit}$ and $|U|_{crit}$ criteria.

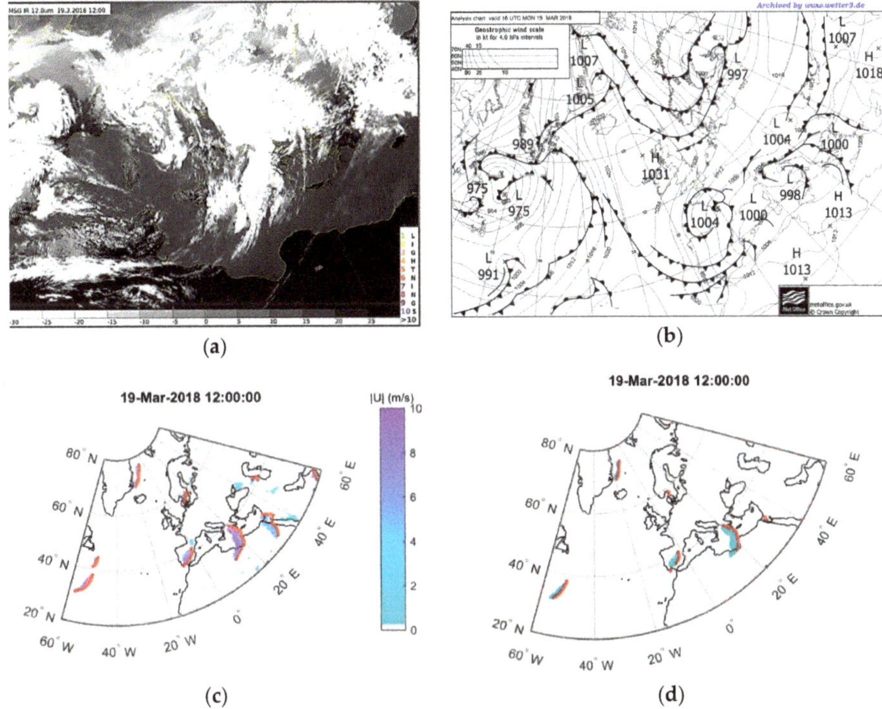

Figure 7. (**a**) Satellite image (IR 12μm) of the Mediterranean sea at 19 March 2018, 12:00UTC, (**b**) synoptic surface chart over the area of interest at the same time, (**c**) identified fronts for $d\varphi_{crit} = 30°$ and $|U|_{crit} = 5$ m s^{-1}. Red lines represent the identified fronts, whereas colored areas show the magnitude of the total wind $|U|$ at the grid points where the $d\varphi$ criterion is met. (**d**) Respective results for the case of solely the dv criterion for $dv_{crit} = 6$ m s^{-1}.

4. Statistical Validation

After the above modifications, the new MedFTS scheme was applied for a ten-year period (2007–2016) in order to validate its ability in identifying Mediterranean cold fronts on climatological basis. The number of the cold fronts passing over Greece was counted for the specific synoptic hour of 00:00 UTC. Then, the results were validated against synoptic analyses obtained from Deutscher Wetterdienst, with the aid of statistical indices (Table 1). It should be noted that the statistical validation handles the occurrence of a cold front as a two-fold categorical variable, and therefore if two fronts appear at the same time over the examined area in the analysis or in the scheme, only one is counted in the total number. Due to the limited geographic area of the examined region, the appearance of more than one front is an extremely rare event and does not affect the obtained results.

The total number of the cold fronts identified by the scheme was $a + b = 511$, which is slightly lower than the corresponding number identified from the charts ($a + c = 547$). From Figure 8 we see that there is excellent agreement between the monthly frontal frequencies in the two datasets, albeit with a scheme underestimation in spring and summer and overestimation in autumn and winter. Furthermore, the scheme succeeds in capturing the intra-annual variation of the frequency of cold fronts, in agreement with the results of [22]. The vast majority of the simulated cold fronts is observed during the cold period of the year, from November to March, peaking in February and March. The frequency declines after April with minimum in August due to the predominance of anticyclonic circulation during summer over the Eastern Mediterranean [26].

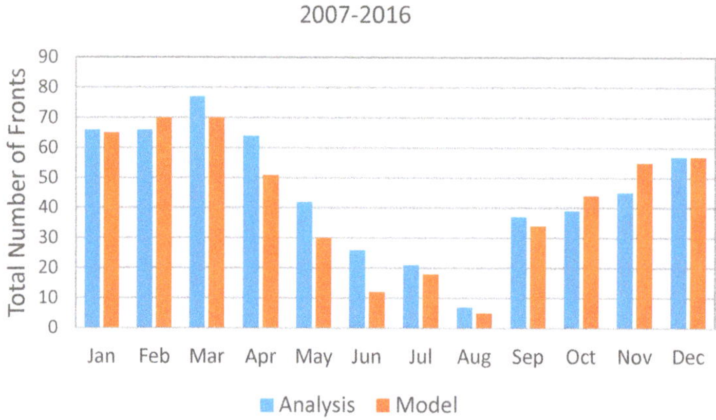

Figure 8. Mean annual cycle of the number of cold fronts identified by the MedFTS (model) and the synoptic charts (analysis) over Greece for the period 2007–2016.

Furthermore, using the indices of Table 1, we calculated the success metrics described in Table 2, taken from [33] and given in Table 3. It can be seen that the scheme succeeded in identifying correctly the bulk of fronts in synoptic charts (hits) while it correctly rejected the vast majority of the fronts that did not appear in the synoptic charts (correct rejection). On the contrary, the number of false alarms and misses are comparatively smaller. It becomes also evident that the false alarms (b) are slightly more than the misses (c).

Table 1. Definition of statistic indices used for the comparison of the fronts identified by the algorithm and the fronts appearing in the synoptic charts.

Symbol	Statistic Index	Explanation
a	hits	Front exists in synoptic charts and identified
b	false alarms	Front identified but not appearing in charts
c	misses	Front appearing in charts not identified
d	correct rejection	No front identified and no front in charts
n	$a+b+c+d$	Sample size

Table 2. Definition of the statistical metrics used of the validation of the algorithm's capability.

Metric	Definition	Range	Perfect Score
Frequency Bias Index (FBI)	$\frac{a+b}{a+c}$	$0 \div \infty$	1
Probability of Detection (POD)	$\frac{a}{a+c}$	$0 \div 1$	1
False Alarm Ratio (FAR)	$\frac{b}{a+b}$	$0 \div 1$	0
Critical Success Index (CSI)	$\frac{a}{a+b+c}$	$0 \div 1$	1
True Skill Statistics (TSS)	$\frac{ad-bc}{(a+c)(b+d)}$	$-1 \div 1$	1
Heidke Skill Score (HSS)	$\frac{2(ad-bc)}{(a+c)(c+d)+(a+b)(b+d)}$	$-\infty \div 1$	1
Equitable Threat Score (ETS)	$\frac{a-a_r}{a+b+c-a_r}$ where $a_r = \frac{(a+b)(a+c)}{n}$	$-1/3 \div 1$	1

Table 3. Values of the indices of Table 1, as they are counted for the decade 2007–2016.

Number of Fronts	Fronts Appearing in Synoptic Charts	Fronts Not Appearing in Synoptic Charts
Fronts appearing in the scheme	$a = 436$	$b = 111$
Fronts not appearing in the scheme	$c = 75$	$d = 3031$

Table 4 gives the values obtained for the metrics of Table 2. It can be seen that the value of FBI is almost equal to the perfect score 1, suggesting that the total number of the fronts in the scheme ($a + b$) is almost equal with the total number of fronts appearing in synoptic charts ($a + c$). Therefore, the scheme is unbiased, without apparent overestimation or underestimation of the front frequency appearing in the charts. Furthermore, POD was found of 85% and FAR of about 20%, suggesting satisfactory detection of the observed fronts with limited false identifications. Besides, Critical Success Index (CSI) presents high value (0.701), taking into account both incorrectly identified fronts and unidentified fronts. Since CSI is somewhat sensitive to the climatology of the event, the Equitable Threat Score (ETS) is also used, providing comparable value.

Table 4. Values of the metrics that are defined in Table 2 for the decade 2007–2016.

Metric	Value
FBI	1.070
POD	0.853
FAR	0.203
CSI	0.701
TSS	0.818
HSS	0.794
ETS	0.659

5. Conclusions

In this study, the University of Melbourne Frontal Tracking Scheme (FTS), a front tracking algorithm based solely on wind related criteria that was employed in Southern Hemisphere, was appropriately modified to identify cold fronts in the Mediterranean, a closed basin with complex topography. It was then used to compile a climatology of these features. The modified scheme (named MedFTS) employs two new criteria, i.e., total wind direction change and total wind magnitude, to better identify the position and tilt of a Mediterranean cold front. Different threshold values of the combined criteria were tested for 20 different cases of cold fronts in the Mediterranean and an optimum selection of critical values was obtained.

It was found that the total wind shift, along with the wind magnitude, accounts for good representation of different types of cold fronts in the Mediterranean throughout the year. Therefore, in order for the Mediterranean cold fronts to be identified, not only the wind shift is required but a sufficient wind magnitude as well. Since no thermodynamic criteria has been included in this scheme, it is implied that wind shift is a prerequisite for the transition of a baroclinic zone to an organised cold front in the Mediterranean, confirming the experience of operational forecasters that, if there is no wind component at the upper levels perpendicular to the low level baroclinic zone, the formation of the frontal zone is inhibited [34]. This wind shift is mostly related with upper level disturbances, approaching a pre-existing area of enhanced low level temperature gradients [35].

Considering the climatological component of the scheme, a statistical validation of its results referring to the frequency of cold fronts passing over Greece for a decade was performed against results derived manually from synoptic analyses. It was found that the total frequency of the identified cold fronts agreed very well with the frequency of the fronts identified from synoptic analyses over Greece. Furthermore, the scheme succeeded in capturing the inter-monthly variations of the frequency of cold fronts. The employment of statistical metrics, considering the front as a two-fold categorical variable, confirms the satisfactory performance of the MedFTS on a climatological basis.

We are exploring the value of including thermodynamic and moisture information into the MedFTS, although we are aware that these are not without problems and weaknesses [36]. However, we have seen that with the appropriate modifications for the Mediterranean region, the dynamically-based FTS can be successfully applied to cold front identification.

Author Contributions: Conceptualization, H.F. and J.K.; methodology, I.S. and I.R.; software, E.B., I.S., I.R.; validation, E.B. and J.K.; formal analysis, E.B., M.H.; investigation, E.B. and M.H.; data curation, M.H., J.K.; writing—original draft preparation, E.B., H.F., M.H.; writing—review and editing, E.B., H.F., M.H.; visualization, E.B.; supervision, H.F.

Funding: This work has been partly funded by the Greek Academy of Athens under a PhD scholarship.

Conflicts of Interest: The authors declare no conflict of interest. The funders had no role in the design of the study; in the collection, analyses, or interpretation of data; in the writing of the manuscript, or in the decision to publish the results.

References

1. Anderson, R.; Boville, B.W.; McClellan, D.E. An operational frontal contour-analysis model. *Q. J. R. Meteorol. Soc.* **1955**, *81*, 588–599. [CrossRef]
2. Catto, J.L.; Pfahl, S. The importance of fronts for extreme precipitation. *J. Geophys. Res. Atmos.* **2013**, *118*, 10791–10801. [CrossRef]
3. Godson, W.L. Synoptic properties of frontal surfaces. *Q. J. R. Meteorol. Soc.* **1951**, *77*, 633–653. [CrossRef]
4. Miller, J.E. On the concept of frontogenesis. *J. Meteorol.* **1948**, *5*, 169–171. [CrossRef]
5. Scherhag, R. *Neue Methoden der Wetteranalyse und Wetterprognose*; Springer-Verlag: Berlin, Heidelberg, 1948; p. 424.
6. Taljaard, J.J.; Schmitt, W.; van Loon, H. Frontal analysis with application to the Southern Hemisphere. *Notos* **1961**, *10*, 25–58.
7. Berry, G.; Reeder, M.J.; Jakob, C. A global climatology of atmospheric fronts. *Geophys. Res. Lett.* **2011**, *38*, L04809. [CrossRef]
8. Hope, P.; Keay, K.; Pook, M.; Catto, J.; Simmonds, I.; Mills, G.; McIntosh, P.; Risbey, J.; Berry, G. A Comparison of Automated Methods of Front Recognition for Climate Studies: A Case Study in Southwest Western Australia. *Mon. Weather Rev.* **2014**, *142*, 343–363. [CrossRef]
9. Hewson, T.D. Objective fronts. *Meteorol. Appl.* **1998**, *5*, 37–65. [CrossRef]
10. Berry, G.; Jakob, C.; Reeder, M. Recent global trends in atmospheric fronts. *Geophys. Res. Lett.* **2011**, *38*, L21812. [CrossRef]
11. Catto, J.; Raveh-Rubin, S. Climatology and dynamics of the link between dry intrusions and cold fronts during winter. Part I: Global climatology. *Clim. Dyn.* **2019**, *53*, 1873–1892. [CrossRef]
12. Renard, R.J.; Clarke, L.C. Experiments in numerical objective frontal analysis. *Mon. Weather Rev.* **1965**, *93*, 547–556. [CrossRef]

13. Clarke, L.C.; Renard, R.J. The U.S. Navy numerical frontal analysis scheme: Further development and limited evaluation. *J. Appl. Meteorol.* **1966**, *5*, 764–777. [CrossRef]
14. Thomsen, G.L.; Reeder, M.J.; Smith, R.K. The diurnal evolution of cold fronts in the Australian subtropics. *Q. J. R. Meteorol. Soc.* **2009**, *135*, 395–411. [CrossRef]
15. Schemm, S.; Rudeva, I.; Simmonds, I. Extratropical fronts in the lower troposphere–global perspectives obtained from two automated methods. *Q. J. R. Meteorol. Soc.* **2015**, *141*, 1686–1698. [CrossRef]
16. Hewson, T.D. Diminutive Frontal Waves—A Link between Fronts and Cyclones. *J. Atmos. Sci.* **2009**, *66*, 116–132. [CrossRef]
17. Mills, G.A. A re-examination of the synoptic and mesoscale meteorology of Ash Wednesday 1983. *Aust. Meteorol. Mag.* **2005**, *54*, 35–55.
18. McIntosh, P.; Pook, M.; Risbey, J.; Hope, P.; Wang, G.; Alves, O. Australia's Regional Climate Drivers. CAR6 Final Rep. *Land Water Aust.* **2008**, 57.
19. Simmonds, I.; Keay, K.; Bye, J.A.T. Identification and climatology of southern hemisphere mobile fronts in a modern reanalysis. *J. Clim.* **2012**, *25*, 1945–1962. [CrossRef]
20. Rudeva, I.; Simmonds, I. Variability and Trends of Global Atmospheric Frontal Activity and Links with Large-Scale Modes of Variability. *J. Clim.* **2015**, *28*, 3311–3330. [CrossRef]
21. Blumen, W. Propagation of fronts and frontogenesis versus frontolysis over orography. *Meteorol. Atmos. Phys.* **1992**, *48*, 37–50. [CrossRef]
22. Flocas, A.A. The annual and seasonal distribution of fronts over central-southern Europe and the Mediterranean. *J. Climatol.* **1984**, *4*, 255–267. [CrossRef]
23. Flocas, H.A.; Simmonds, I.; Kouroutzoglou, J.; Keay, K.; Hatzaki, M.; Asimakopoulos, D.N.; Bricolas, V. On cyclonic tracks over the Eastern Mediterranean. *J. Clim.* **2010**, *23*, 5243–5257. [CrossRef]
24. Campins, J.; Genovés, A.; Jansà, A.; Guijarro, J.A.; Ramis, C. A catalogue and a classification of surface cyclones for the Western Mediterranean. *Int. J. Climatol.* **2000**, *20*, 969–984. [CrossRef]
25. Kouroutzoglou, J.; Flocas, H.A.; Keay, K.; Simmonds, I.; Hatzaki, M. Climatological aspects of explosive cyclones in the Mediterranean. *Int. J. Climatol.* **2011**, *31*, 1785–1802. [CrossRef]
26. Hatzaki, M.; Flocas, H.A.; Simmonds, I.; Kouroutzoglou, J.; Keay, K.; Rudeva, I. Seasonal aspects of an objective climatology of anticyclones affecting the Mediterranean. *J. Clim.* **2014**, *27*, 9272–9289. [CrossRef]
27. Parfitt, R.; Czaja, A.; Seo, H. A simple diagnostic for the detection of atmospheric fronts. *Geophys. Res. Lett.* **2017**, *44*, 4351–4358. [CrossRef]
28. Dal Piva, E.; Gan, M.A.; Rao, V.B. Energetics of winter troughs entering South America. *Mon. Weather Rev.* **2010**, *138*, 1084–1103. [CrossRef]
29. McAndrew, A. *An Introduction to Digital Image Processing with MATLAB*; Course Technology Press: Boston, MA, USA, 2004; p. 528.
30. Velleman, P.F.; Hoaglin, D.C. *Applications, Basics and Computing of Exploratory Data Analysis*; Duxbury Press: Boston, MA, USA, 1981.
31. Jenkner, J.; Sprenger, M.; Schwenk, I.; Schwierz, C.; Dierer, S.; Leuenberger, D. Detection and climatology of fronts in a high-resolution model reanalysis over the Alps. *Meteorol. Appl.* **2010**, *17*, 1–18. [CrossRef]
32. Dee, D.P.; Uppala, S.M.; Simmons, A.J.; Berrisford, P.; Poli, P.; Kobayashi, S.; Andrae, U.; Balmaseda, M.A.; Balsamo, G.; Bauer, P.; et al. The ERA-Interim reanalysis: Configuration and performance of the data assimilation system. *Q. J. R. Meteorol. Soc.* **2011**, *137*, 553–597. [CrossRef]
33. Wilks, D. *Statistical Methods in the Atmospheric Sciences*, 2nd ed.; International Geophysics Series; Academic Press: Cambridge, MA, USA, 2006; Volume 91, Chapter 7 (Forecast Verification).
34. Bader, M.J.; Forbes, G.S.; Grant, J.R.; Lilley, R.B.E.; Waters, A.J. *Images in Weather Forecasting: A Practical Guide for Interpreting Satellite and Radar Imagery*; Cambridge University Press: Cambridge, UK, 1995.
35. Sanders, F. Real front or baroclinic trough? *Weather Forecast.* **2005**, *20*, 647–651. [CrossRef]
36. Rudeva, I.; Simmonds, I.; Crock, D.; Boschat, G. Midlatitude fronts and variability in the Southern Hemisphere tropical width. *J. Clim.* **2019**, *32*, 8243–8260. [CrossRef]

Article

Modeling the Effects of Anthropogenic Land Cover Changes to the Main Hydrometeorological Factors in a Regional Watershed, Central Greece

Angeliki Mentzafou [1,2,*], George Varlas [1], Elias Dimitriou [1], Anastasios Papadopoulos [1], Ioannis Pytharoulis [3] and Petros Katsafados [2]

1 Institute of Marine Biological Resources and Inland Waters, Hellenic Centre for Marine Research, 19013 Anavissos, Greece; gvarlas@hcmr.gr (G.V.); elias@hcmr.gr (E.D.); tpapa@hcmr.gr (A.P.)

2 Department of Geography, Harokopio University of Athens, 17671 Kallithea, Greece; pkatsaf@hua.gr

3 Department of Meteorology and Climatology, Aristotle University of Thessaloniki, 54124 Thessaloniki, Greece; pyth@auth.gr

* Correspondence: angment@hcmr.gr; Tel.: +30-2291076349

Received: 24 September 2019; Accepted: 6 November 2019; Published: 7 November 2019

Abstract: In this study, the physically-based hydrological model MIKE SHE was employed to investigate the effects of anthropogenic land cover changes to the hydrological cycle components of a regional watershed in Central Greece. Three case studies based on the land cover of the years 1960, 1990, and 2018 were examined. Copernicus Climate Change Service E-OBS gridded meteorological data for 45 hydrological years were used as forcing for the model. Evaluation against observational data yielded sufficient quality for daily air temperature and precipitation. Simulation results demonstrated that the climatic variabilities primarily in precipitation and secondarily in air temperature affected basin-averaged annual actual evapotranspiration and average annual river discharge. Nevertheless, land cover effects can locally outflank the impact of climatic variability as indicated by the low interannual variabilities of differences in annual actual evapotranspiration among case studies. The transition from forest to pastures or agricultural land reduced annual actual evapotranspiration and increased average annual river discharge while intensifying the vulnerability to hydrometeorological-related hazards such as droughts or floods. Hence, the quantitative assessment of land cover effects presented in this study can contribute to the design and implementation of successful land cover and climate change mitigation and adaptation policies.

Keywords: anthropogenic land cover changes; hydrological model MIKE-SHE; time-series statistical analysis; trend analysis; Spercheios river basin

1. Introduction

The quantitative and qualitative state of water resources of a watershed are formed by a variety of drivers that interact in a complex and often indirect way [1]. Climate elements (precipitation; relative humidity; wind speed and direction; solar radiation and temperature that also controls evaporation/evapotranspiration and snow melt and their temporal and spatial distribution) and the biogeophysical characteristics of a catchment (topography, land—vegetation cover, geological structure, soil coverage) are fundamental determinants of regional hydrology [2,3]. The conceptual model describing the interactions among the abovementioned drivers is the hydrological cycle that links the exchange, storage and movement of water among the biosphere, atmosphere, cryosphere, lithosphere, anthroposphere, and hydrosphere [4], while the quantification of the relationships among the components of the hydrological cycle at a given location constitutes the water balance [5].

All characteristics of the catchment (climate elements and biogeophysical characteristics) are factors that can be largely affected by anthropogenic activities and pressures [3]. Humanly imposed climate change due to increased emissions of greenhouse gases and dust from anthropogenically-disturbed soils [6], is expected to significantly increase freshwater-related risks such as modification of the hydrological regime, floods and droughts, and to affect water cycle components [7,8]. Climate change is projected to reduce renewable surface water and groundwater resources significantly in most dry subtropical regions and is likely to increase the frequency of meteorological droughts (less rainfall) and agricultural droughts (less soil moisture) in presently dry regions. Additionally, projections imply variations in the frequency of floods and negative impacts on freshwater ecosystems by changing streamflow and water quality [4,7]. Regarding anthropogenic interventions in catchment's physical characteristics (for example alteration of the land surface soil moisture, albedo and roughness [9]), land cover changes due to livestock grazing, agriculture, timber harvest, deforestation, and urbanization can reduce retention of water in watersheds and lead to an increase of the size and frequency of floods and to the reduction of baseflow levels [10]. Dam constructions and diversion, canalization, snagging and dredging of rivers, streams and drainage ditches, and groundwater overexploitation, disrupt the dynamic equilibrium between the movement of water and sediment that exists in rivers [10]. Based on recent studies, direct human impacts on the terrestrial water cycle are in some large river basins of the same order of magnitude, or even larger than climate change [11,12]. Especially land cover change alters annual global runoff to a similar or greater extent than other major drivers [13], while land use change contribution in regional runoff values in tropical regions is larger than that of climate change [14], especially in the case of smaller catchments [15].

Worldwide studies support the impacts of land cover changes, mainly deforestation and urbanization, on the hydrometeorological factors, leading primary to river discharge increase [16–23] and generally to an increase of eco-environmental vulnerability of the watersheds [24,25]. In Greece, studies confirm the impact of land cover change and deforestation in river discharge. For example, a study conducted in Pinios river basin proved that expanding the agricultural land over forest by 20%, a mean monthly increase in the river discharge of up to 3%, can be observed from October to April and a respective reduction from May to September, reaching a maximum of 6% in July [26]. Moreover, human interference in streams crossing urban or suburban areas raise the vulnerability to flash floods. For example, the hydrometeorological analysis of a fatal flash flood event which occurred on 15 November 2017 in the suburban area of Mandra, western Attica, Greece resulting in extensive damages and 24 fatalities, showcased heavy storm-induced run-off water in combination with human pressures on streams as the reason for the flood [27].

Regarding the impact of land cover change to evapotranspiration, it has been reported that mean annual evapotranspiration can be up to 39% lower in agricultural ecosystems than in natural ecosystems in Brazil [20]. A recent study concerning the whole of China showed that the average annual land surface evapotranspiration decreased at a rate of −0.6 mm/yr from 2001 to 2013, attributed partly to land use and land cover changes of forests to other land types [28]. In Greece, a study in a small catchment showed that 16% increase of agricultural land against wetland and forest area led to a 6% increase of evapotranspiration and 10% increase of the water deficit in the soil [29].

Given the uncertainty of future land cover changes due to socio-economic driving forces and local development policies applied, a scenario-based modeling framework can be beneficial in supporting the analysis of potential land cover changes, so as to mitigate potentially negative future impacts on a basin's water resources. In order to investigate the effect of anthropogenic land cover changes to the hydrological cycle components and the main hydrometeorological factors of a regional agricultural watershed in Central Greece (Spercheios river basin) of great ecological value, three (3) land cover case studies were adopted, based on the land cover distribution documented in the following years: in 1960 (hereafter LC1960; baseline), in 1990 (hereafter LC1990; mid-period), and in 2018 (hereafter LC2018; current state). The modeling tool used was the physically-based hydrological model (MIKE SHE), while the high-resolution gridded observational daily meteorological dataset of Europe named E-OBS [27] from the EU-FP6 project UERRA [30] and the Copernicus Climate Change Service [31] was also employed to drive the model. Since the E-OBS gridded dataset had not been used before in similar studies in Greece, the statistical evaluation of its efficiency was considered to be obligatory before performing any further analysis. Finally, statistical tests and trend analysis were performed on the simulated time series of each land cover case study examined.

The main objective of the present study was the better understanding of the system's response and the basin's water resources to possible future land cover changes, while the main research questions intended to be addressed are: (a) which are the interrelationships among land cover and the main hydrometeorological factors' (precipitation, air temperature, discharge, and actual evapotranspiration) variations, (b) how land cover changes affect the trend magnitude of the main hydrometeorological factors, and (c) which are the hydrometeorological-related hazards associated with land cover changes in the study area?

2. Materials and Methods

2.1. Study Area

Spercheios river basin is located in the prefecture of Pthiotida in Central Greece, covers an area of 1,661 km^2, and has a mean altitude of 641 m and a dense hydrographic network (Figure 1; [32]). The main human activities of the wider area since it was first inhabited in the Early Neolithic period [33] include arable agriculture and grazing, while industrial activities are limited mainly to small manufacturing units of agricultural products and olive oil refineries [34]. The main hydromorphological modifications of the area include water abstractions for irrigation, water flow regulations (small weirs, water distributor), canalization, and the partial diversion of the original route of the river close to its estuary. Spercheios river wider area has been included in many environmental protection networks (for example NATURA 2000, CORINE biotopes, and Wildlife Refuges; [32,35]).

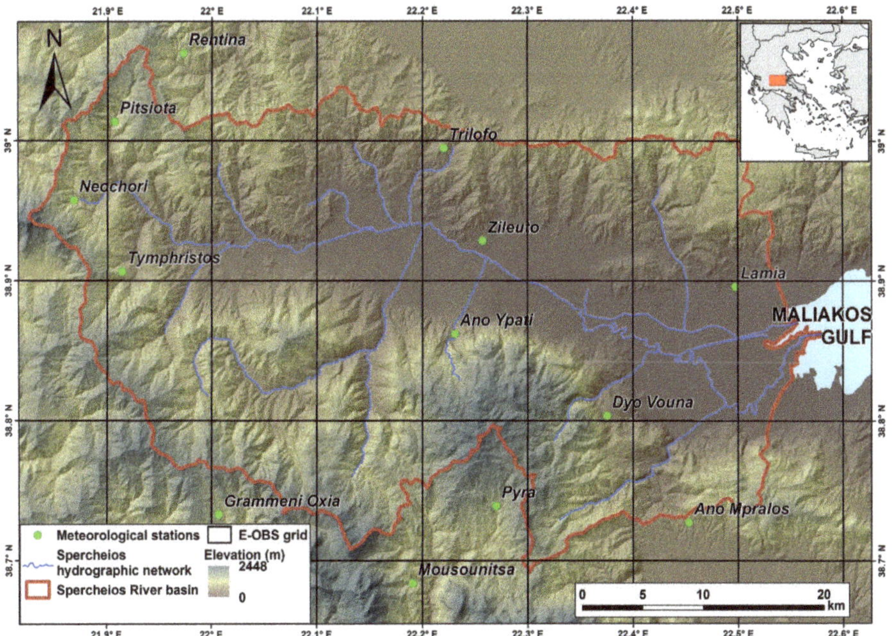

Figure 1. Study area.

2.2. *Hydrological Simulation*

2.2.1. Methodological Approach

The modelling tool used in the present study was the MIKE SHE, developed by the Danish Hydraulic Institute Water and Environment. MIKE SHE is a physically-based distributed model that is able to simulate all hydrological domains within the land phase of the hydrological cycle in a river basin. MIKE SHE is fully integrated with the channel flow code MIKE 11, which is a one-dimensional model that can simulate water flow and level, water quality and sediment transport in rivers, flood plains, irrigation canals, reservoirs, and other inland water bodies [36]. The hydrological model has already been successfully set up, calibrated and validated during a previous study for Spercheios river basin [35].

More specifically, during a previous research study, the hydrological model of Spercheios river basin was set up and calibrated for the hydrological years 2008/2009–2010/2011 and validated for the hydrological years 2013/2014–2014/2015 [35]. These periods were chosen based on the data availability (actual in situ observations of water level and discharge and high-quality climatological data) and on the fact that in 2008 the construction of the last engineering flood control structures in the hydrological network and the river banks were completed. The calibration and validation periods' length were considered to be adequate since most studies addressing the question of the utility of additional data in terms of the length of available discharge time-series in hydrological model calibration concluded that several years of data ranging between 2 and 8 years are sufficient for reliable parameter identification [37]. Moreover, when 2–3 years of continuous daily discharge data are available, then the model activates the complete set of its procedures, and the use of longer data sets would not offer a significant benefit in the definition of the model's uncertainty [38]. The results of the Spercheios river basin hydrological model calibration showed a satisfactory agreement between observed and simulated water levels and discharge measurements. Their correlation coefficient *R* can be characterized as moderate (0.55) to high (0.77) based on the criteria for correlation interpretation proposed by Hinkle et al. [39], and in all cases the data were statistically significant at the 0.05 level, indicating the sufficient performance of the model. During validation, the resulted correlation coefficient *R* was also moderate (0.68) to high (0.84) [39], and the data were also statistically significant at the 0.05 level. The model performance can be considered satisfactory since the results meet the criteria proposed by Moriasi [40] (*R2* > 0.50, *RSR* < 0.70, and *PBIAS* ± 25% for streamflow) in the cases where river discharge data were available for validation [35] (Table 1).

Table 1. Statistical characteristics and efficiency criteria for the calibration and validation of the hydrological model at the Spercheios river basin [35].

Station		Para-Meter	Statistical Parameter								
			N	*ME*	*MAE*	*RMSE*	*R*	*p*-Value	*R2*	*PBIAS*	*RSR*
Kastri bridge	Calibration	L	1095	0.065	0.177	0.238	0.77	<0.00001 *	0.56	-	-
Kompotades bridge	Calibration	L	909	0.042	0.273	0.387	0.55	<0.00001 *	0.11	-	-
Komma bridge	Calibration	L	1063	−0.105	0.351	0.479	0.61	<0.00001 *	0.34	-	-
KR2	Validation	L	176	0.063	0.102	0.138	0.70	<0.00001 *	-1.06	-	-
KR3	Validation	L	236	0.009	0.085	0.143	0.80	<0.00001 *	0.62	-	-
KR6	Validation	L	303	0.146	0.266	0.366	0.68	<0.00001 *	-1.24	-	-
KR7	Validation	L	462	−0.060	0.195	0.405	0.76	<0.00001 *	−0.88	-	-
KR2	Validation	Q	12	0.735	1.164	1.777	0.79	0.002333 *	0.48	32%	0.21
KR3	Validation	Q	11	−0.718	1.304	2.007	0.84	0.001277 *	0.56	−27%	0.20

* result significant at $p < 0.05$; L: Level (m); Q: Discharge (m^3/s); *N*: number of sample pairs; *ME*: mean error; *MAE*: mean absolute error; *RMSE*: root mean squared error; *R*: correlation coefficient; *R2*: Nash-Sutcliffe coefficient of efficiency; *PBIAS*: percent bias; *RSR*: RMSE-observations standard deviation ratio.

In order to investigate the impact of land cover change on the hydrological cycle components, the calibrated hydrological model of Spercheios river basin was integrated for 45 hydrological years (1960/61–2004/05) for three different land cover case studies. The specific period was characterized by a stable hydrographic network with minimum engineering interventions. Any hydraulic construction built after 2006 was omitted from simulation procedure, while all engineering interventions which took place before 1960 were included into the simulation, for the best representation of the actual state of the river network during the period 1960/61–2004/05. For the specific simulation period, the gridded time-series of the meteorological dataset used for the specific study area was complete and without gaps. Finally, the land cover case studies were selected according to the data availability and taking into consideration the overall anthropogenic interventions in the area, aiming at the better representation of each distinguished period. More specifically, until 1960, the major hydraulic interventions and the major agricultural reform of Greece had been completed, and ever since the area used for agricultural activities has been practically stable in the study area [41]. The first available documentation concerning the land cover distribution in Greece was from the year 1960 [42]. In 1990, the first pan-European land cover data collection was utilized based on satellite image processing (Coordination of Information on the Environment- CORINE Land Cover Programme [43]), and the most recent version is from the year 2018 [44]. Therefore, the following three land cover case studies in Spercheios river basin were implemented: (1) LC1960 based on the land cover of Spercheios river basin in 1960 (baseline), (2) LC1990 based on the land cover in 1990 (mid-period) and (3) LC2018 based on the land cover in 2018 (current state) (Figure 2).

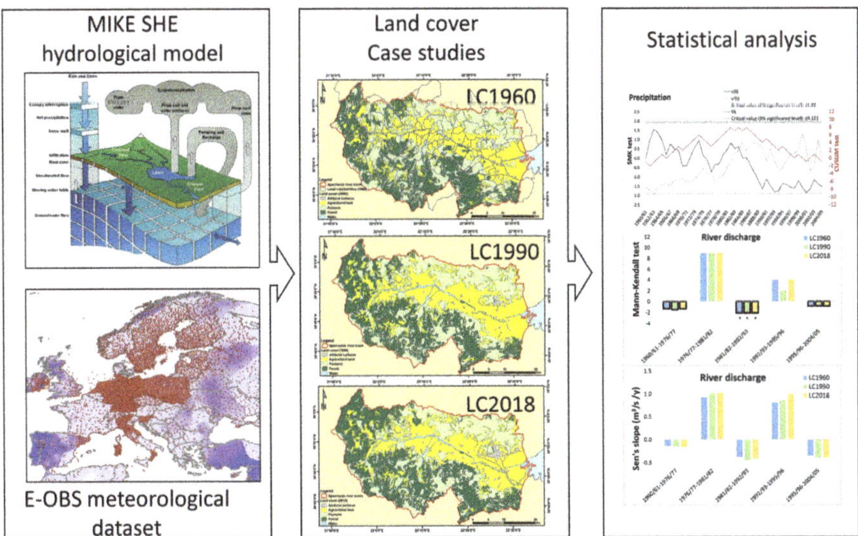

Figure 2. Flowchart of the current methodological approach.

2.2.2. Meteorological Data

Due to the lack of detailed and evenly distributed meteorological time-series in Spercheios river basin covering the entire simulated period (1960/61–2004/05), the necessary data for the hydrological modeling concerning daily temperature (minimum, average and maximum values) and precipitation were retrieved from the high-resolution gridded data set of daily climate over Europe termed E-OBS from Copernicus Climate Change Service [45,46]. The specific data set covers the period back to 1950 and provides high-resolution gridded fields at a spacing of $0.1° \times 0.1°$ in regular latitude/longitude coordinates. The ensemble version of E-OBS v.19.0e (the dataset produced from averaging multiple equally probable interpolations of station-based observations, so as to provide the best representation of the spatial and temporal distribution of climate parameters and a measure of uncertainty [47]) is based on the European Climate Assessment and Dataset (ECA&D) initiative that combines collation of daily series of observations at meteorological stations, quality control, analysis of extremes, and dissemination of both the daily data and the analysis results (Figure 1; [48,49]).

The reliability of the E-OBS dataset was evaluated by comparing the time-series of in-situ observations from meteorological stations installed in the wider area by various agencies (Table 2; Figure 1) against the corresponding grid point of the E-OBS dataset. The statistical criteria used to investigate the dataset reliability were the following: mean error *ME*; mean absolute error *MAE*; root mean squared error *RMSE*; standard deviation *STDEV*; and correlation coefficient *R*, while also the *p*-value was calculated to estimate the significance of the results.

Table 2. Meteorological stations used in estimation of E-OBS efficiency.

Station	Longitude (dd)	Latitude (dd)	Altitude (m)	Owner	Observations Available
Ano_Mpralos	22.45474	38.73054	580.5	MEE	P
Ano_Ypati	22.23273	38.86562	286.0	MEE	P
Dyo Vouna	22.37684	38.80680	470.6	PPC	P
Gr. Oxia	22.00846	38.73601	1107.1	PPC	P, Tmin, Tmax
Lamia	22.49940	38.89895	144.0	HNMS	P, Tmin, Tmax, Tav
Mousounitsa	22.19244	38.68742	846.1	MEE	P, Tmin, Tmax
Neochori	21.86983	38.96068	821.6	PPC	P
Pitsiota	21.90836	39.01663	783.9	PPC	P
Pyra	22.27196	38.74262	1137.1	MEE	P
Rentina	21.97414	39.06577	884.9	MEE	P
Trilofo	22.22209	38.99834	575.3	MEE	P
Tymphristos	21.91575	38.90961	847.9	MEE	P
Zileuto	22.25904	38.93192	97.2	MEE	P

MEE: Ministry of Environment and Energy of Greece; HNMS: Hellenic National Meteorological Service; PPC: Public Power Corporation S.A.; P: precipitation; Tmin: minimum air temperature; Tmax: maximum air temperature; Tav: mean air temperature.

The lack of the necessary climatological data (relative humidity, solar radiation and wind speed) precluded the use of the Penman-Monteith equation for the estimation of daily reference evapotranspiration *ET*. Therefore, *ET* was estimated using the Hargreaves empirical approach [50], which is recommended only in cases of lack of other meteorological data and is considered to provide satisfactory results with an error rate of 10–15% or 1 mm/d, whichever is greater [51,52]. In the Hargreaves approach, except for daily average, minimum and maximum temperature, all the other required parameters (solar radiation, latent heat of vaporization) can be estimated using empirical relationships [52].

2.2.3. Land Cover Spatial Distribution

The oldest official and most detailed information concerning land cover distribution in Spercheios river basin was available from National Statistical Service of Greece for the year 1960. These data were part of the preparatory activities taken place prior the Agricultural and Livestock Census of March 19, 1961 [42] and concerned the main land cover types per local community: agricultural land, communal or private pastures for grazing animals, forest, artificial surfaces and water. It should be noted that in the 1960s land cover census, all agricultural activities (annual, crops, vineyards, tree plantations, and fallow land) were grouped together, while areas covered by shrubs, transitional woodland—shrub areas or areas with dense vegetation were characterized as pastures. During this procedure, forests were defined as areas mainly covered by ligneous plants clearly supported by a trunk and branching out to no less than 1 m from the ground. The category artificial surfaces included cities, settlements, roads, mines, and bare rocks. Finally, the category water included lakes, permanent inland and salt marshes, coastal areas and lagoons, estuaries, water courses, river beds, and areas covered by water for the greatest part of the year. Areas temporarily covered by water and areas lying near rivers or lakes dried and usually cultivated in summer were included in arable land (Table 3).

Table 3. Land cover nomenclature used in the present study.

Level 1	Level 2	1960	1990	2018
Artificial surfaces	Continuous urban fabric	x	x	x
	Discontinuous urban fabric	x	x	x
	Airports		x	x
	Bare rocks		x	
	Industrial or commercial units		x	x
	Mineral extraction sites		x	x
	Construction sites			x
	Road and rail networks and associated land			x
	Sport and leisure facilities			x
Agricultural land	Non-irrigated arable land	x	x	x
	Permanently irrigated land	x	x	x
	Complex cultivation patterns	x	x	x
	Land principally occupied by agriculture	x	x	x
	Rice fields	x	x	x
	Olive groves	x	x	x
	Vineyards	x	x	
	Fruit trees and berry plantations			x
Pastures	Natural grasslands	x	x	x
	Sclerophyllous vegetation	x	x	x
	Transitional woodland-shrub	x	x	x
	Pastures	x	x	x
	Moors and heathland		x	x
	Sparsely vegetated areas		x	x
Forest	Broad-leaved forest	x	x	x
	Coniferous forest	x	x	x
	Mixed forest	x	x	x
Water	Sea and ocean		x	x
	Estuaries		x	x
	Beaches, dunes, sands		x	x
	Salt marshes	x	x	x
	Water courses	x	x	x

The production of the 1960 land cover map (LC1960) was carried out by distributing the land uses per local community, by also taking into consideration the land cover distribution of the CORINE Land Cover (CLC) inventory for the year 1990 [44] and the Census of Agricultural and Livestock Holdings for the year 1961 [53]. As mentioned above, detailed agricultural activities were not distinct in 1960s land cover documented distribution [42]; therefore, the estimation of the different agricultural classes was based on their corresponding distribution per local community for the year 1990 [44]. Natural grasslands, pastures, sclerophyllous vegetation, transitional woodland-shrub, moors and heathland, and sparsely vegetated areas were classified as pastures, while artificial surfaces included continuous and discontinuous urban fabric, airports, industrial or commercial units, mineral extraction sites, construction sites, and road and rail networks (Table 3). The spatial distribution of forest classes (broad-leaved, coniferous and mixed) was also based on the CORINE Land Cover (CLC) inventory for the year 1990 [44].

The land cover maps for the years 1990 (LC1990) and 2018 (LC2018) were retrieved from CORINE Land Cover (CLC) inventory for the corresponding years [44]. It should be noted that based on the methodological approach of CORINE Land Cover (CLC), the density of houses is the main criterion to attribute a land cover class to the discontinuous urban fabric or to the agricultural area, in complex cultivation patterns class. In case of patchwork of small agricultural parcels and scattered houses, the cut-off-point to be applied for discontinuous urban fabric is 30% at least of urban fabric within the patchwork area [43]. Therefore, documented sparsely populated areas in 1960 and 1990 land cover distributions were in many cases classified as complex cultivation patterns (Table 3; Figure 3).

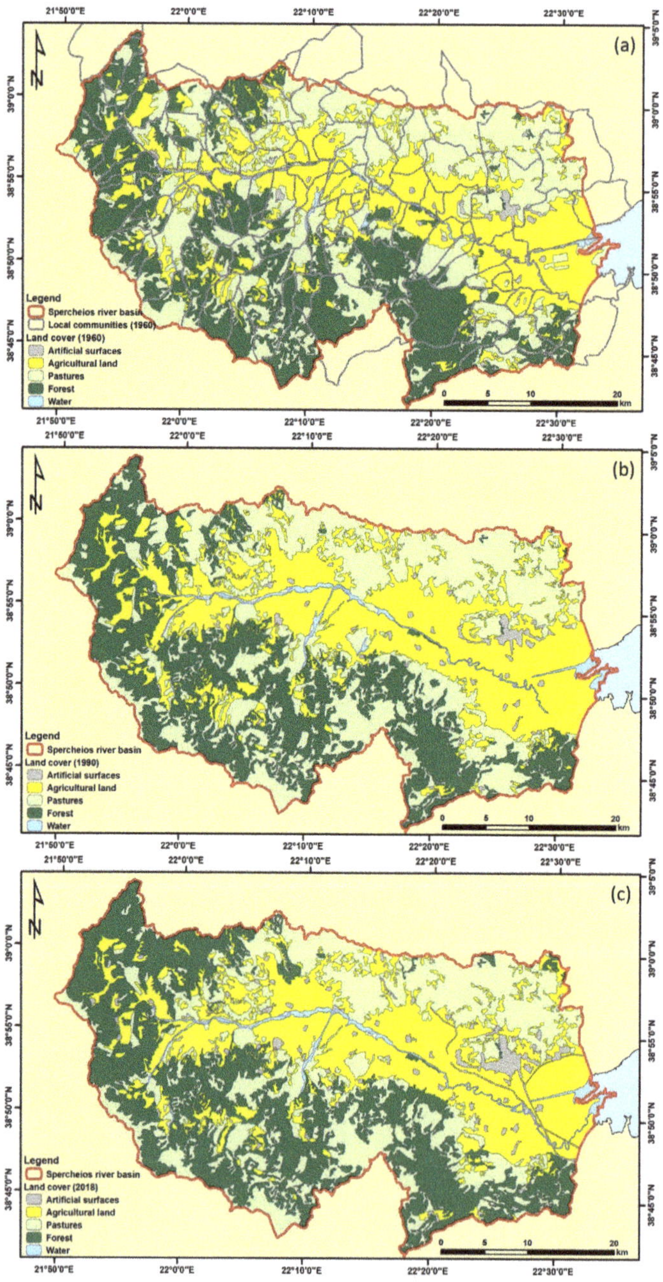

Figure 3. Land cover maps for the year: (**a**) 1960 produced for the present study, (**b**) 1990, and (**c**) 2018 from CORINE Land Cover (CLC).

2.2.4. Irrigation Demands

In order to estimate the irrigation demands for each land cover case study in LC1960, LC1990 and LC2018, due to a lack of detailed information concerning the crops cultivated for each reference year, the irrigation demands for the year 2010 were used. More specifically, during a previous study [35], the irrigation demands for Spercheios river basin per local community were calculated taking into consideration the detailed agricultural activities and the cultivated crops per local community from the Census of Agricultural and Livestock Holdings 2010 [54] and the methodology proposed by Food and Agriculture Organization of the United Nations (FAO) for the estimation of the net irrigation requirement of each crop in the study area [55]. The irrigation demands were then projected for the years 1960 (LC1960), 1990 (LC1990) and 2018 (LC2018) per local community, based on the ratio of the corresponding irrigation area as documented during the corresponding census or annual agricultural statistical surveys [53,56,57] and the irrigation area of 2010 (for Pthiotida prefecture; Table 4).

Table 4. Estimation of irrigation demands for Spercheios river basin.

Year	Reference Year [1]	Total Cultivated Area (km^2) [2]	Total Irrigated Area (km^2) [2]	Ratio of Irrigated Area [2]	Ratio of Irrigation Area (in Relation to 2010) [2]	Total Annual Irrigation Demand (*10^6 m^3)
2010	2009	1388	573	41%	-	129.8 [35]
1960	1961	1416	279	20%	49%	63.3
1990	1990	1459	520	36%	91%	117.9
2018	2017	1176	480	41%	84%	108.8

[1] Year when Census of Agricultural and Livestock Holdings was conducted; [2] Pthiotida prefecture.

2.3. Statistical Analysis

In order to investigate the effects of anthropogenic land cover changes to the annual actual evapotranspiration and river discharge, the following statistical tests were applied to the time-series of each land cover case study examined.

The sequential version of the Mann-Kendall (SMK) test [58] and the non-parametric rank-based distribution-free cumulative sum CUSUM test [59] were applied, so as to detect the approximate change of trend with time. The sequential version of the Mann-Kendall test (SMK) is calculated so that rank (x_i) > rank (x_j) ($i > j$). The number of cases $x_i > x_j$ is counted and denoted by n_i. The t statistic is calculated as Equation (1):

$$t = \sum_{i=1}^{n} n_i \tag{1}$$

The distribution of t is assumed to be asymptotically normal with the following expectations (Equations (2) and (3)):

$$E(t) = \mu = \frac{n(n-1)}{4} \tag{2}$$

and

$$Var(t) = \sigma^2 = \frac{n(n-1)(2n+5)}{72}. \tag{3}$$

The null hypothesis that there is no trend is rejected for high values of the reduced variable $|u(t)|$, which is calculated as Equation (4):

$$u(t) = \frac{t - E(t)}{\sqrt{Var(t)}}. \tag{4}$$

Similar to the calculation of the sequential progressive series $u(t)$, the retrograde series $u'(t)$ is computed backwards starting from the end of the time-series [58]. The intersection of the curves $u(t)$ and $u'(t)$ indicates the approximate turning point of the trend of the original time-series. For the trend to be significant, the point of intersection must exceed the critical values of the confidence level. The sign of the curve $u(t)$ indicates whether the trend is increasing or decreasing.

In CUSUM test, the test statistic V_k is defined as Equation (5):

$$V_k = \sum_{i=1}^{k} sgn(x_i - x_{median}),\ \mathrm{k} = 1,\ 2,\ \ldots,\ \mathrm{n} \tag{5}$$

Where x_{median} the median value of the x_i data set and $sgn(x)$. CUSUM test allows the detection of changes in mean value of a sequence of observations ordered in time, by comparing successive observations with the median of the series. If a significant trend develops in the plotted points either upward or downward, it is evidence that the process mean has shifted and further investigation is required [59–61].

After estimating the approximate change of trend with time, the rank-based non-parametric Mann-Kendall test [62,63] was applied to each sub-period, so as to identify the trend significance. The Mann-Kendall statistic S compares each value of the series (x_t) with all subsequent values (x_{t+1}) and is defined as Equation (6):

$$S = \sum_{t'=1}^{n-1} \sum_{t=t'+1}^{n} \mathrm{sgn}(x_t - x_{t'}) \tag{6}$$

where sgn is the sign function (Equation (7)).

$$\mathrm{sgn}(x_t - x_{t'}) = \begin{cases} 1,\ if\ x_t > x_{t'} \\ 0,\ if\ x_t = x_{t'} \\ -1,\ if\ x_t < x_{t'} \end{cases} \tag{7}$$

If $n < 10$, the absolute value of S is compared directly to the theoretical distribution of S derived by Mann and Kendall [64]. When $n \geq 10$, the statistic S is approximately normally distributed with the mean m and the variance V as follows [62,63] (Equation (8)).

$$ES = 0,\ V(S) = \frac{1}{18}\left[n(n-1)(2n+5) - \sum\nolimits_{i=1}^{g} e_i(e_i - 1)(2e_i + 5)\right] \tag{8}$$

g is the number of tied groups, and e_i is the number of data in the ith tied group. The values of S and $VAR(S)$ are used to compute the test statistic Z. The standardized test statistic Z is defined as follows (Equation (9)).

$$Z = \frac{S + m}{\sqrt{V(S)}}, \tag{9}$$

Finally, the trend magnitude for each trend period identified from the abovementioned statistical test was calculated based on Sen's estimator of slope. This non-parametric statistic can be applied in cases of linear trend and determines the magnitude of change per unit time [65]. The Sen's slope estimation test is defined for a season g as follows (Equation (10)):

$$\beta = Median\left(\frac{x_i - x_j}{i - j}\right),\ \mathrm{i} < \mathrm{j} \tag{10}$$

where β; the slope between points x_i and x_j, x_i data measurement at time i, and x_j data measurement at time j. The positive value of the β; implies the slope of the upward trend and negative value for the downward trend [66].

3. Results

3.1. Long-Term Land Cover Changes

Based on the results, the land cover of Spercheios river basin has changed considerably over the last five decades. The artificial surfaces have increased during the years reaching from 1% in 1960 and 1990 to 3% of the total river basin area in 2018. This can be partly attributed to the fact that in some cases, small settlements in 1990 were classified as agricultural areas due to the 30% threshold adopted in the methodology by European Environmental Agency in CLC inventory for distinguishing discontinuous urban fabric and complex cultivation patterns [43].

Agricultural land ranges from 28% (470 km^2) in 1960, through 32% (531 km^2) in 1990, to 30% (498 km^2) in 2018. Permanently irrigated land has increased from 2% (40 km^2) in 1960, through 4% (71 km^2) in 1990, to 8% (135 km^2) in 2018. On the contrary, non-irrigated land has decreased from 12% (196 km^2) in 1960, through 10% (171 km^2) in 1990, to 5% (79 km^2) in 2018. Other agricultural activities, the majority of which were also irrigated, range from 14% (235 km^2) in 1960, through 17% (289 km^2) in 1990, to 17% (284 km^2) in 2018. Pastures have decreased over the last decades (from 636 km^2—38% in 1960, through 599 km^2—36% in 1990, to 538 km^2—32% in 2018). Finally, forested land change ranges from 31% (517 km^2) in 1960, through 30% (492 km^2) in 1990, to 34% (558 km^2) in 2018 (Figures 3 and 4).

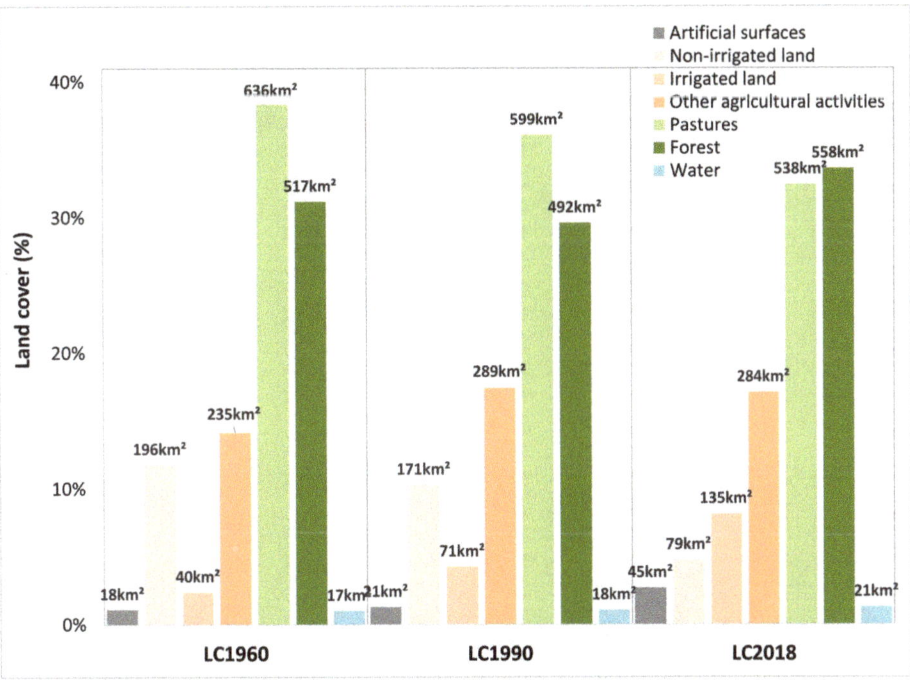

Figure 4. Distribution of land cover type for the three land cover case studies examined.

Figure 5 presents the differences in agricultural land (Figure 5a), pastures (Figure 5b) and forests (Figure 5c) for the three land cover case studies examined. Some of the areas characterized by transitions in land cover among the cases studies were used to investigate the impact of land cover change on annual actual evapotranspiration.

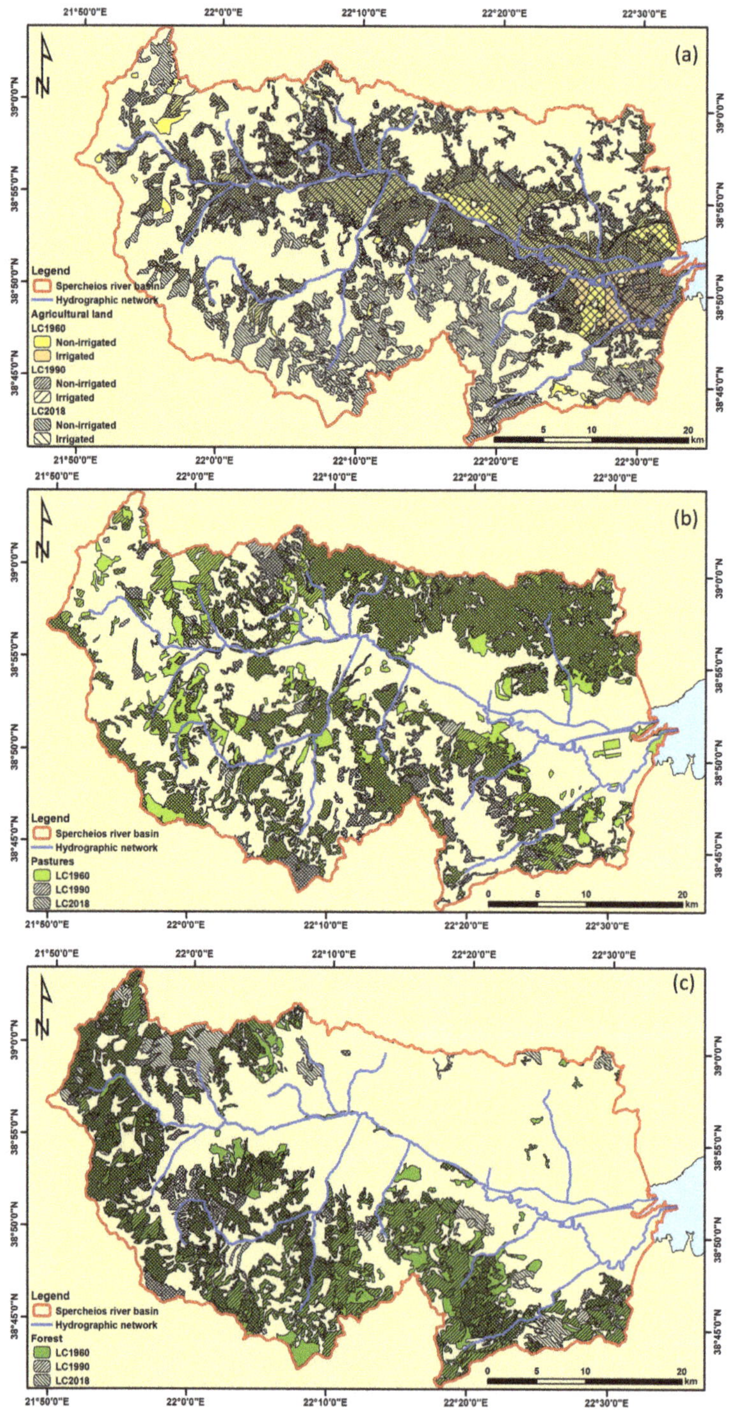

Figure 5. Differences in agricultural land (**a**), pastures (**b**) and forests (**c**) for the three land cover case studies examined.

3.2. Meteorological Data

Based on the results of the comparison between in-situ observations from ground meteorological stations and the E-OBS dataset, there was sufficient agreement regarding precipitation (Table 5). The correlation coefficient *R* ranged between low (0.3) to very high positive (0.9; based on the criteria for correlation interpretation proposed by Hinkle et al. [39]); nevertheless, the *p*-value in all cases was statistically significant at the 0.05 level, except in the case of the meteorological station Ano Mpralos. Overall, the E-OBS dataset systematically underestimated annual precipitation for the entire period of evaluation, except in the case of Zilefto meteorological station for which the *ME* was calculated to be positive. E-OBS dataset was not able to sufficiently estimate the altitude effect on the precipitation rate, leading to a higher value of *ME* in meteorological stations of higher elevation (Table 5; Figure 6). This led to an average 37% underestimation of spatially-averaged annual precipitation of Spercheios river basin.

Table 5. Statistical characteristics and efficient criteria of annual observed precipitation measurements and E-OBS dataset.

Station	Station Altitude (m)	Period	*N*	*AV* (mm)		*ME* (mm)	*MAE* (mm)	*RMSE* (mm)	*R*	*p*-Value	Significance		
				Station	E-OBS						$p < 0.01$	$p < 0.05$	$p < 0.10$
AMpr	580.5	1970/71–2004/05	35	996.9	489.8	−507.1	517.2	645.6	0.38	0.1033	No	No	No
AYp	286	1960/61–2004/05	45	727.0	584.1	−142.8	214.5	281.1	0.4	0.0184	No	Yes	Yes
DVou	470.6	1980/81–2000/01	21	1024.6	473.2	−551.4	551.4	571.8	0.9	<0.00001	Yes	Yes	Yes
GrOx	1107.1	1980/81–2000/01	21	1059.8	598.3	−461.5	461.5	484.4	0.5	0.0157	No	Yes	Yes
Lam	144	1970/71–2004/05	35	569.0	471.7	−97.29	102.4	122.6	0.8	<0.00001	Yes	Yes	Yes
Mous	846.1	1963/64–2004/05	42	1228.1	644.8	−583.2	613.0	680.2	0.4	0.0131	No	Yes	Yes
Neo	821.6	1960/61–1991/92	32	1671.6	674.5	−997.1	997.1	1022.9	0.8	<0.00001	Yes	Yes	Yes
Pits	783.9	1960/61–1991/92	32	1257.0	635.3	−621.7	621.7	637.7	0.8	<0.00001	Yes	Yes	Yes
Pyr	1137.1	1963/64–2004/05	39	1070.6	628.4	−442.2	470.0	514.9	0.4	0.0063	Yes	Yes	Yes
Rent	884.9	1960/61–2004/05	45	1246.5	617.6	−628.9	628.9	745.1	0.7	<0.00001	Yes	Yes	Yes
Tril	575.3	1960/61–2004/05	45	528.5	481.0	−47.49	87.3	110.8	0.7	<0.00001	Yes	Yes	Yes
Tymf	847.9	1960/61–2004/05	45	989.1	638.7	−350.3	389.6	446.8	0.4	0.0052	Yes	Yes	Yes
Zil	97.2	1960/61–2003/04	41	481.1	487.6	6.5	98.1	128.6	0.5	0.0025	Yes	Yes	Yes

N: number of observations; *AV*: Average; *ME*: mean error (=E-OBS − station observed values); *MAE*: mean absolute error; *RMSE*: root mean square error; *R*: correlation coefficient; AMpr: Ano Mpralos; AYp: Ano Ypati; DVou: Dyo Vouna; GrOx: Grammenni Oxia; Lam: Lamia; Mous: Mousounitsa; Neo: Neochori; Pits: Pitsiota; Pyr: Pyra; Rent: Rentina; Tril: Trilofo; Tymf: Tymphristos; Zil: Zileuto.

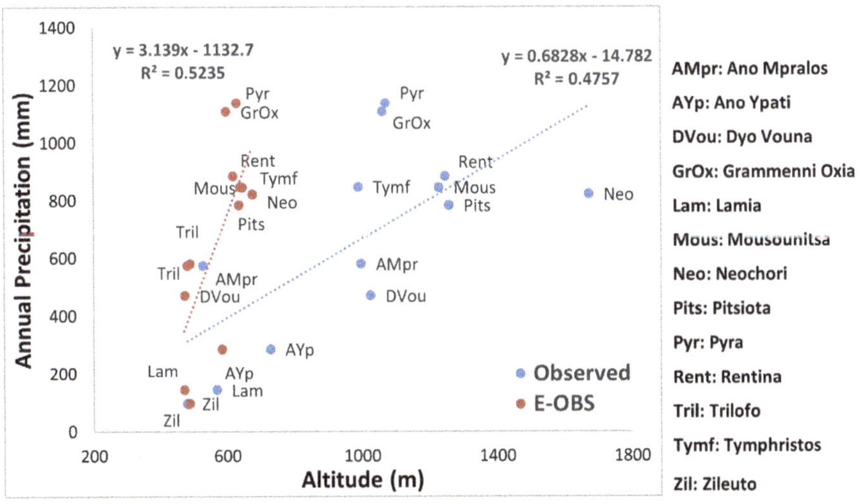

Figure 6. Precipitation lapse rates of Spercheios river basin based on observational meteorological data and E-OBS dataset.

Concerning air temperature, the E-OBS dataset managed to represent the actual measurements efficiently, except in the case of minimum air temperature, based on the higher *MAE* statistics calculated in minimum temperature in all stations (and maximum temperature of Mousounitsa station). The correlation coefficient R ranged between moderate (0.49) to very high (0.98) positive [39]; nevertheless, the p-value was not statistically significant at the 0.10 level in the case of minimum temperature at Lamia station and at 0.05 level in the case of maximum temperature at Mousounitsa station. Overall, temperature was underestimated (ME negative in all cases), especially in the case of minimum air temperature of Lamia station and of minimum and maximum air temperature of Mousounitsa station (Table 6).

Table 6. Statistical characteristics and efficient criteria of annual observed air temperature measurements and E-OBS dataset.

Station	Period	N	AV (°C)		ME (°C)	MAE (°C)	RMSE (°C)	R	p-Value	Significance		
			Station	E-OBS						$p < 0.01$	$p < 0.05$	$p < 0.10$
Gr.Oxia (Tmin)	1973/74–1996/96	24	6.3	3.9	−2.4	2.4	2.5	0.46	0.0237	No	Yes	Yes
Gr.Oxia (Tmax)	1973/74–1996/96	24	15.5	15.1	−0.4	0.9	1.1	0.49	0.0151	No	Yes	Yes
Lamia (Tmin)	1970/71–2003/04	34	11.1	7.5	−3.5	3.5	3.7	0.10	0.5736	No	No	No
Lamia (Tmax)	1970/71–2003/04	34	21.9	19.4	−2.5	2.5	2.5	0.98	<0.00001	Yes	Yes	Yes
Lamia (Tav)	1970/71–2003/04	34	16.6	13.9	−2.8	2.8	2.8	0.84	<0.00001	Yes	Yes	Yes
Mousounitsa (Tmin)	1993/94–2004/05	12	9.0	4.0	−5.0	5.0	5.4	0.58	0.0465	No	Yes	Yes
Mousounitsa (Tmax)	1993/94–2004/05	12	20.3	14.9	−5.4	5.4	6.2	0.53	0.0763	No	No	Yes

Tmin: minimum temperature; Tmax: maximum temperature; Tav: average temperature; N: number of observations; AV: Average; ME: mean error (=E-OBS − station observed values); MAE: mean absolute error; $RMSE$: root mean square error; R: correlation coefficient.

3.3. Statistical Analysis

Based on the descriptive statistics of the simulated time-series of the main hydrometeorological factors at Spercheios river basin, the mean annual precipitation of the entire catchment for the period 1960/61–2004/05 was 542.5 mm, and the mean annual air temperature was 13.2 °C for the same period. Mean annual river discharge to Maliakos Gulf ranged from 5.1 m^3/s in LC1960, through 5.7 m^3/s in LC1990, to 5.4 m^3/s in LC2018, while annual actual basin-averaged evapotranspiration ranged from 406.1 mm in LC1960, through 384.7 mm in LC1990, to 395.0 mm in LC2018 (Table 7; Figure 7). Hence, in comparison with LC1960, LC1990 and LC2018 case studies estimated 11.8% and 5.9% higher mean annual river discharge to Maliakos Gulf, respectively. On the other hand, they estimated 5.3% and 2.5% lower basin-averaged annual actual evapotranspiration. These results can be attributed to water balance which force discharge and actual evapotranspiration to be "communicating vessels", given the same meteorological forcing in the three land cover case studies examined. It is interesting to note that the results showcased the role of richly-vegetated area variabilities on the hydrological characteristics of the catchment. Deforestation as well as intertemporal increase of artificial surfaces have negative effects on evapotranspiration while increasing discharge. For example, the reduced forested area of LC1990 in comparison with both LC1960 and LC2018, resulted in minimum basin-averaged annual actual evapotranspiration and maximum mean annual river discharge.

Table 7. Descriptive statistics for the annual time-series of the main hydrometeorological factors.

Factor		Minimum	Maximum	Mean	Std. Deviation	Variance
Precipitation (mm)		315.6	912.4	542.5	109.1	11,902.0
Air temperature (°C)		12.2	14.8	13.2	0.6	0.4
Basin-averaged actual evapotranspiration (mm)	LC1960	298.1	515.5	406.1	47.9	2291.4
	LC1990	288.9	488.9	384.7	46.3	2141.7
	LC2018	305.7	496.0	395.0	44.9	2020.2
River discharge to Maliakos Gulf (m^3/s)	LC1960	1.1	14.4	5.1	2.5	6.2
	LC1990	1.3	15.4	5.7	2.7	7.3
	LC2018	1.1	15.1	5.4	2.7	7.2

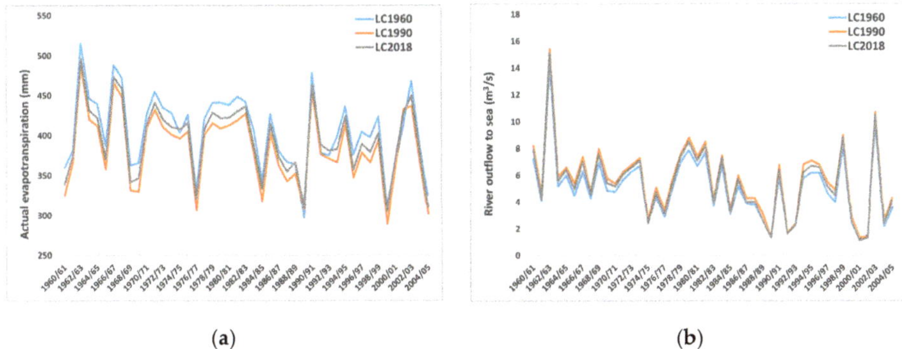

(**a**) (**b**)

Figure 7. Basin-averaged annual actual evapotranspiration (mm) (**a**) and river discharge (m^3/s) to Maliakos Gulf (**b**) reconstructions. LC1960, LC1990 and LC2018 are represented by blue, orange and gray lines.

As far as mean annual actual evapotranspiration is concerned, the three land cover case studies present spatial differences. LC1960 is characterized by more inhomogeneous patterns than in LC1990 and LC2018 (Figure 8). It resulted in values exceeding 520 mm/yr and lower than 280 mm/yr at many areas. A possible explanation for this difference is the increased scattering of areas covered by forests, agricultural land and pastures in LC1960 which have effects of different magnitude on evaporation and transpiration (see Figure 2). Forests and agricultural land increased evapotranspiration in comparison with pastures and artificial surfaces [67]. In contrast to LC1960, LC1990 and LC2018 case studies are characterized by wider continuous areas covered by the same land cover. Hence, the pattern of mean annual actual evapotranspiration is smoother in LC1990 and LC2018 than in LC1960.

Comparing the results of the three land cover case studies, annual actual evapotranspiration is almost the same at areas covered by artificial surfaces over time, for example at the city of Lamia (not shown). On the other hand, the transition from pastures to agricultural land or forest increased mean annual actual evapotranspiration, while the inverse transition had the opposite effects. In order to quantitively estimate the impact of each land cover transition, case studies are compared in pairs e.g. LC1960 vs LC 1990, regarding spatially averaged actual evapotranspiration only at areas characterized by a specific transition. As far as deforestation is concerned, the transition from LC1960 to pastures in LC1990 decreased annual actual evapotranspiration with a mean rate of about 33 mm/yr (Figure 9a,b). It is important to note that the reduction is also evidenced for the entire simulation period which means that land cover change effects can locally outflank the impact of climatic variability [11,12]. However, the transition from pastures in LC1960 to agricultural land in LC2018 increased annual actual evapotranspiration (Figure 9c,d) with a mean rate of about 24 mm/yr. On the other hand, the transition from forest in LC1990 to agricultural land in LC2018 caused a reduction in annual actual evapotranspiration with a mean rate of about 26 mm/yr. Although, differences exist at areas with the same land cover in all three cases examined, these are quite a bit smaller than those which appeared at areas where land cover changed. These smaller differences may be attributed to the horizontal propagation of land cover effects. The local differences in land cover introduce complex forcing in parameters such as run-off water, infiltration, evaporation, and transpiration which can sharply affect in a non-linear way the spatial distribution of water balance, yielding local differences in evapotranspiration.

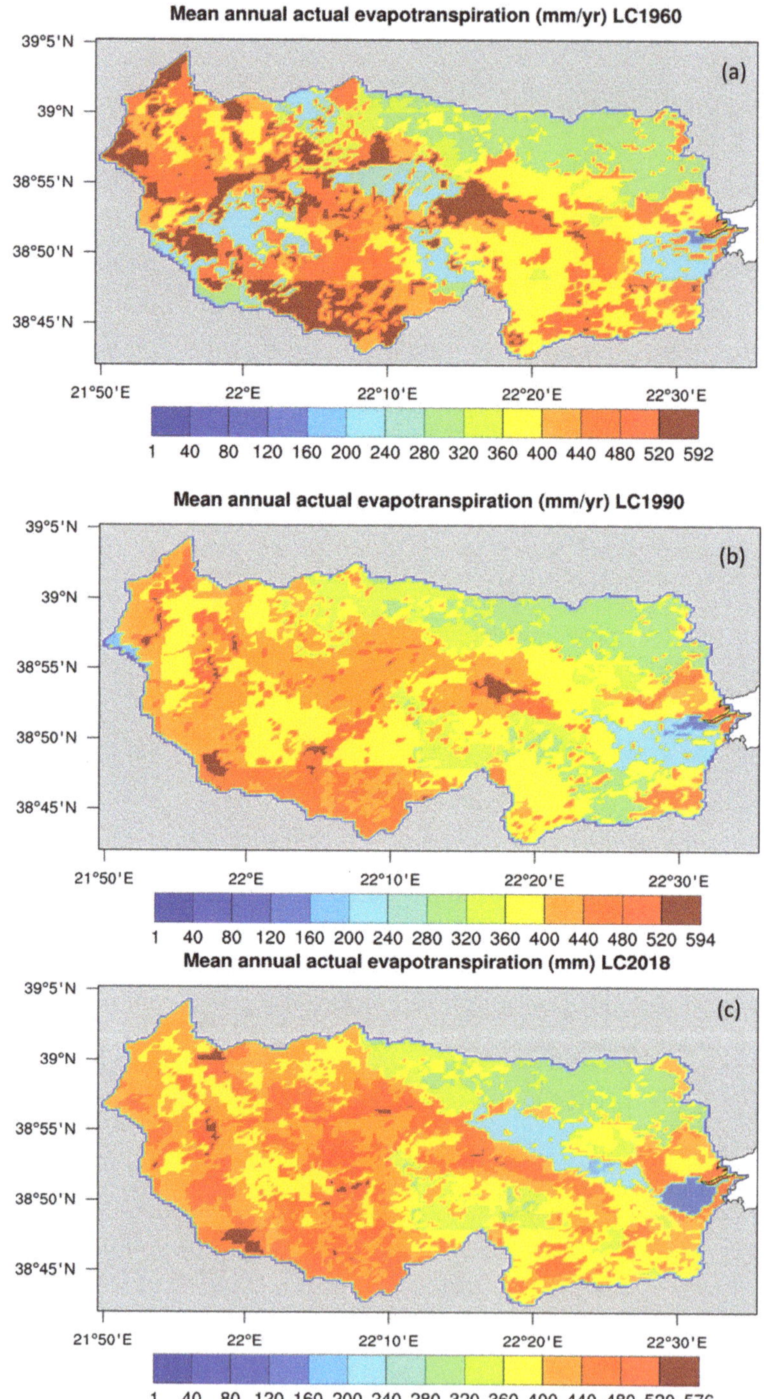

Figure 8. Simulated mean annual actual evapotranspiration (mm/yr) in (**a**) 1960, (**b**) 1990 and (**c**) 2018.

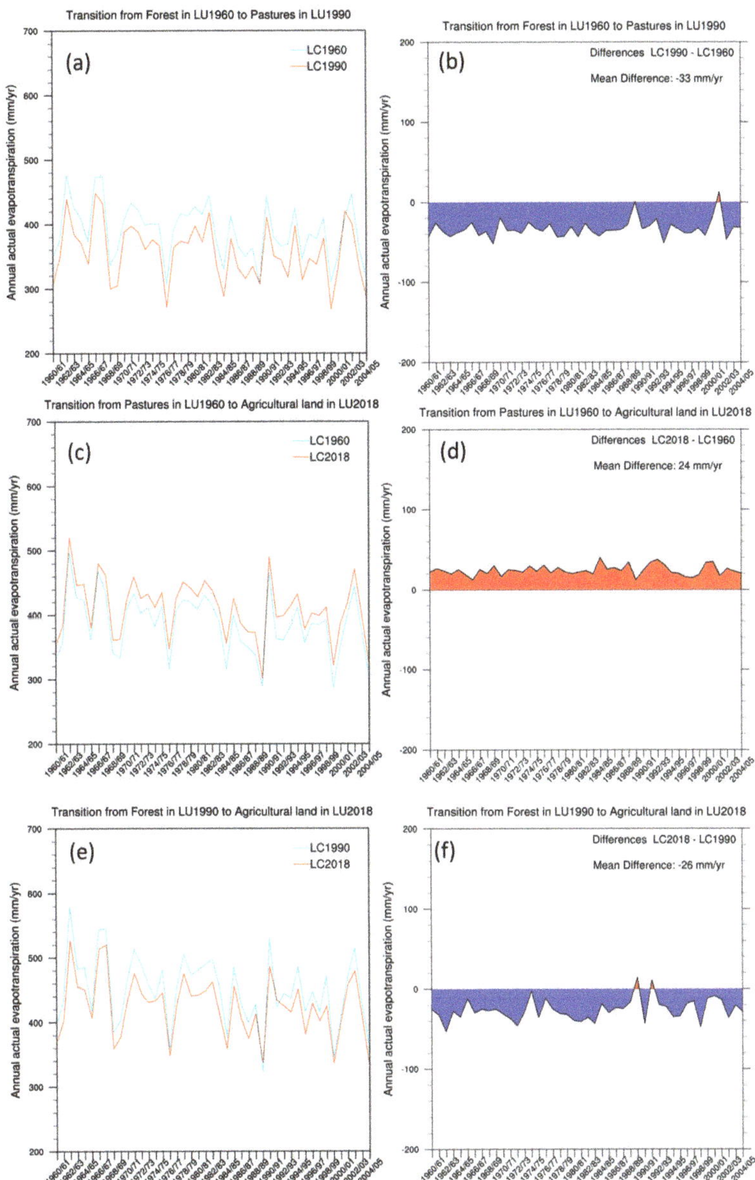

Figure 9. Timeseries, regarding areas characterized by transition from forest in 1960 to pastures in 1990, of (**a**) annual actual evapotranspiration (mm/yr) simulated by LC1990 (orange line) and LC1960 (blue line) as well as (**b**) their differences (red for positive and blue for negative). The same for (**c**,**d**) as well as (**e**,**f**) regarding transition from pastures in 1960 to agricultural land in 2018 and transition from forest in 1990 to agricultural land in 2018, respectively.

The statistical tests applied on the time-series of the main hydrometeorological factors (precipitation, air temperature, actual evapotranspiration, and river discharge) for the trend analysis and change point detection, resulted in the following findings. Although the $u(t)$ and $u'(t)$ curves of precipitation intersect only at one point (1981/82), the following trends were identified based on the general form of the $u(t)$ curve. C, concerning annual precipitation: (1) three increasing periods were identified (1960/61–1973/74, 1976/77–1981/82 and 1992/93–2002/03), and (2) four decreasing periods (1973/74–1976/77, 1981/82–1992/93 and 2002/03–2004/05 respectively). It should be noted that all trends identified, either with SMK test, either with CUSUM test, were not significant at the 0.05 confidence level (Figures 10a and 11a).

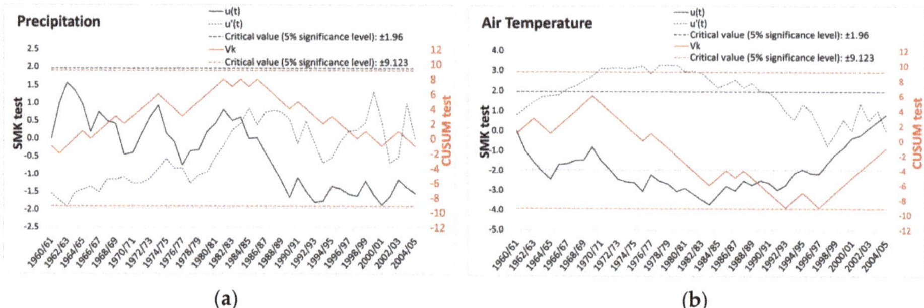

Figure 10. SMK and CUSUM tests for (**a**) precipitation and (**b**) air temperature.

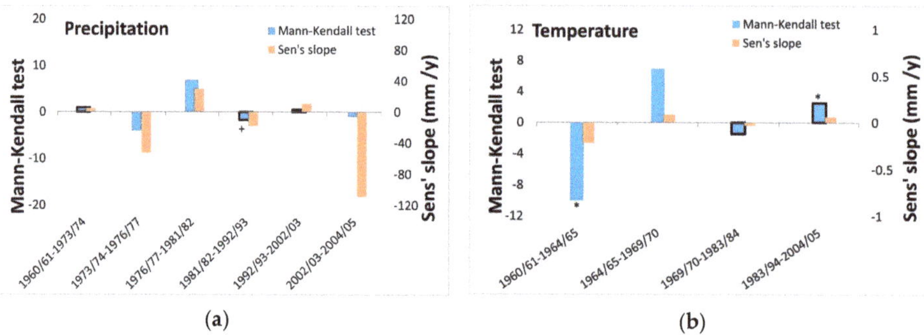

Figure 11. Results of Mann-Kendall test (No border in columns indicates S test and black border Z test; (*) and (+) symbols indicate if trend is significant at $\alpha = 0.05$ level and at $\alpha = 0.1$ level respectively) and Sen's slope for (**a**) precipitation and (**b**) air temperature.

Regarding annual air temperature, the following trends were identified, based on the form of the $u(t)$ curve: (a) two decreasing periods were identified (1960/61–1964/65 and 1969/70–1983/84), followed (b) by two increasing periods (1964/65–1969/70 and 1983/94–2004/05 respectively). Of the abovementioned trend periods, based on the SMK test, the periods 1960/61–1964/65 and 1983/94–2004/05 were significant at the 0.05 confidence level, while with the CUSUM test, the trends identified were not statistically significant at the 0.05 confidence level (Figures 10b and 11b).

The trend analysis and change point detection tests applied in actual evapotranspiration time-series of all land cover case studies examined, led to the identification of: (1) three increasing periods (1960/61–1967/68, 1977/78–1982/83 and 1989/90–1994/95), followed by (2) three decreasing periods (1967/68–1977/78, 1982/83–1989/90 and 1994/95–2004/05 respectively; Figure 12a–c). Nevertheless, the trend magnitude of each period was different for each land cover case study examined. More specifically, the trend magnitude in all trend periods identified was higher in the case of 1960, followed by the trend magnitude calculated for the periods 1990 and 2018, with the exception of the period 1977/78–1982/83 that the trend magnitude was greater in LC2018, followed by LC1990 and LC1960, and the period 1982/83–1989/90, where trend magnitude was practically identical in all land cover cases examined. It should be noted that all trends identified were not significant at the 0.05 confidence level with the SMK test, while with the CUSUM test, the period 1982/83–1989/90 was statistically significant at the 0.05 confidence level in all land cover case studies examined (Figures 13a and 14a).

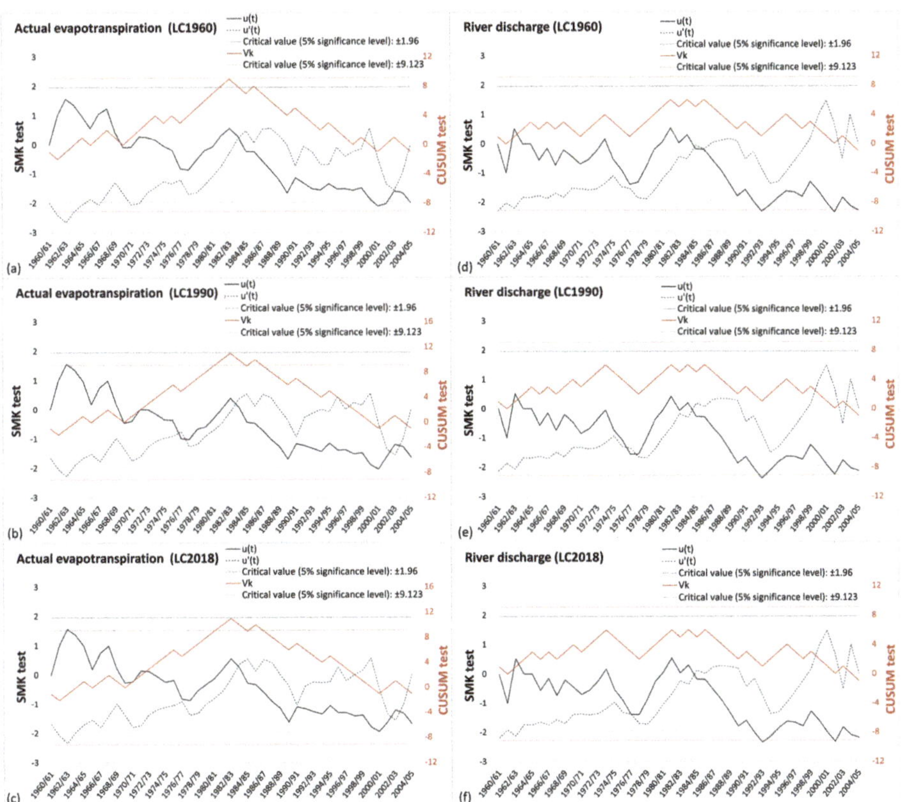

Figure 12. SMK and CUSUM tests for actual evapotranspiration for (**a**) LC1960, (**b**) LC1990, and (**c**) LC2018, and river outflow to Maliakos Gulf for (**d**) LC1960, (**e**) LC1990, and (**f**) LC2018.

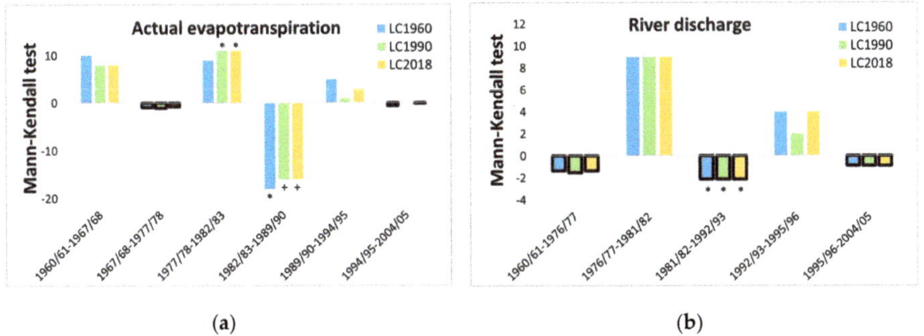

Figure 13. Results of Mann-Kendall test for actual evapotranspiration (**a**) and river discharge (**b**) for LC1960, LC1990, and LC2018. No border in columns indicates S test and black border Z test. (*) and (+) symbols indicate if the trend is significant at the $\alpha = 0.05$ level and $\alpha = 0.1$ level, respectively.

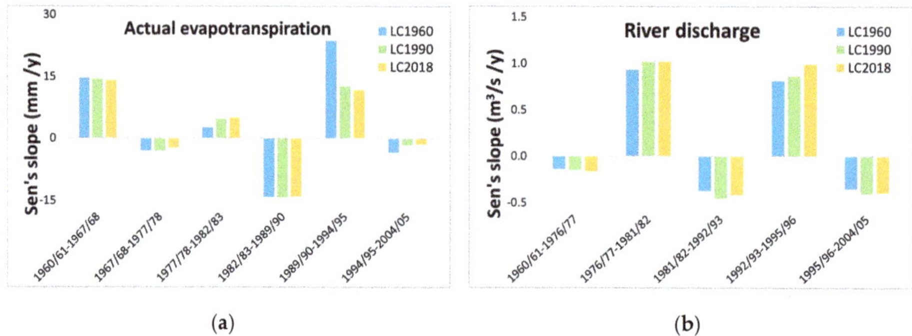

Figure 14. Sen's slope for actual evapotranspiration (**a**) and river discharge (**b**) for LC1960, LC1990, and LC2018.

Finally, concerning Spercheios annual river discharge and outflow to Maliakos Gulf, three decreasing periods were identified for all land cover case studies examined (1960/61–1976/77, 1981/82–1992/93 and 1995/96–2004/05), followed by two increasing periods (1976/77–1981/82 and 1992/93–1995/96; Figure 12d–f). In all trend periods identified, the trend magnitude was smaller in LC1960, followed by LC1990 and LC2018, except in the case of the period 1981/82–1992/93 that the trend magnitude for LC2018 was smaller than in the case of LC1990. These results revealed a significant impact of land cover on the formation of extreme hydrometeorological events. This finding indicates that the decrease of a richly-vegetated area, for example due to deforestation between LC1960 and 1990, increased annual river discharge while intensifying the vulnerability to extreme climatic variabilities which often provokes either droughts or floods. Of the abovementioned trend periods, based on the SMK test, 1981/82–1992/93 was significant at the 0.05 confidence level, while with the CUSUM test, the trends identified were not statistically significant at the 0.05 confidence level (Figures 13b and 14b).

3.4. Water Budgets

Regarding the water budgets of each land cover case study examined, the following can be stated. The actual evapotranspiration at Spercheios river basin ranged in the three land cover case studies examined from 74.9% (LC1960), through 70.9% (LC1990), to 72.8% (LC2018). Baseflow to river ranged from 16.7% (LC1960), through 19.2% (LC1990), to 18.1% (LC2018). Storage change ranged from 11.6% (LC1960), through 13.3% (LC1990), to 12.4% (LC2018) (Figure 15).

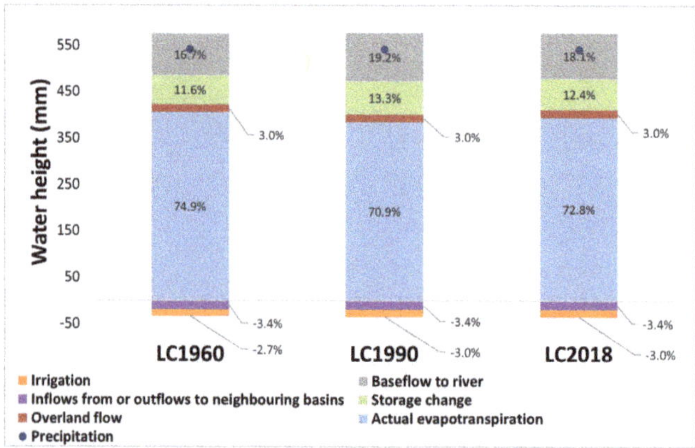

Figure 15. Water budget of Spercheios river basin for the three land cover case studies examined.

4. Discussion

Anthropogenic land cover changes and interventions on catchment's characteristics can be leading factors affecting the hydrological cycle components and, in some cases, the impacts can be of the same order of magnitude, or even larger than those attributed to climatic variabilities [11,12]. In order to investigate the effects of land cover changes on the main hydrometeorological factors of a regional river basin in Central Greece, a physically-based hydrological model (MIKE SHE) and gridded observational meteorological data (Copernicus Climate Change Service E-OBS) were employed, and three land cover case studies were adopted.

Before the simulations, the reliability of the E-OBS dataset including precipitation and daily temperature (average, minimum and maximum) was evaluated by comparing against time-series of in-situ observations from meteorological stations at the basin. Based on the results, E-OBS dataset systematically underestimated precipitation in Spercheios river basin for the entire period of evaluation. This may be attributed to issues arising in the comparison of in-situ measurements with area-averaged estimates [68], such as the identification of the most representative grid-point for each meteorological station, the insufficient density of the weather stations network in Spercheios river basin or possible uncertainties concerning the accuracy of observational measurements [69]. Moreover, the coarse horizontal resolution of E-OBS prevented to accurately describe the influence of topography on precipitation and to adequately resolve the atmospheric mesoscale processes; 10–15 km grid spacing of meteorological variables generally improves the realism of the results but does not necessarily significantly improve the objectively scored accuracy of the forecasts [70]. Additionally, the coarse network of Greek meteorological stations used in the E-OBS development that are not evenly distributed and do not cover higher altitude sufficiently, eventually does not allow the accurate representation of area-averaged estimates. More specifically, the spatially-averaged annual precipitation calculated at the present study for the period 1960/61–2004/05 was 542.5 mm, which is close to the mean annual precipitation of Lamia meteorological station (585.5 mm for the period 1970–2000 [71]). In other studies, the spatially-averaged annual precipitation of Spercheios river basin was estimated to be 836 mm for the period 2008/09–2010/11 (precipitation estimated based on Thiessen polygons method [72]) and 1,077 mm for the wet hydrological years 2013/14–2014/15 [73] (simulated precipitation provided by Poseidon Monitoring, Forecasting and Information System [74]) [75], while for the period 1949/50–1989/90, the spatially-averaged annual precipitation for Spercheios river basin was estimated to be 904.6 mm [76]. Nevertheless, the main scope of the present study was the trend analysis of the time-series of the main hydrometeorological factors and, therefore, these discrepancies were considered to be acceptable, since no other meteorological data except from the low-altitude Lamia meteorological station (Hellenic National Meteorological Service, WMO 16675) were available for the entire simulation period (1960/61–2004/05).

As far as the results of hydrological simulations are concerned, average annual actual evapotranspiration and river discharge were the main parameters of the hydrological cycle which were analyzed in this study. First, the average annual actual evapotranspiration at Spercheios river basin was −5.3% and −2.5% decrease in LC1990 and LC2018 respectively, in comparison to LC1960. These variations can be attributed to the presence of the larger areas covered by vegetation (forest and pastures) in LC1960 (70% in comparison to 66% in LC1990 and LC2018), and especially to the larger extent of areas classified as pastures that also include shrubs, transitional woodland—shrub areas or areas with dense vegetation (38% in LC1960), that led to increased actual evapotranspiration. The higher value of actual evapotranspiration in LC2018 in comparison to LC1990 can be attributed to the increased forested land (34% in LC2018 in comparison to 30% in LC1990). The simulations also presented high spatial differences in average annual actual evapotranspiration. Land cover in 1960 was characterized by a more inhomogeneous pattern than in 1990 and in 2018 due to the increased distribution patterns of areas covered by forests, agricultural land and pastures in 1960 which have different effects on evaporation and transpiration. Moreover, mean annual actual evapotranspiration was almost the same at areas covered by artificial surfaces over time, for example Lamia city, but presents variations where land cover changed. The transition from pastures to agricultural land or forest increased evapotranspiration, while the inverse transition had the opposite effects for the entire simulation period which means that land cover effects can locally outflank the impact of climatic variability.

Second, average annual river discharge to Maliakos Gulf was +11.8% and +5.9% increased in LC1990 and LC2018 respectively, in comparison to LC1960. This can partially be attributed to the contribution of the baseflow to river, that ranged from 16.7% in LC1960, through 19.2% in LC1990, to 18.1% in LC2018, following the same pattern. Additionally, the high forested land covering the area of Spercheios river watershed in the case of LC1960 (31%) combined with the lowest irrigation demands during the same period and led to the smallest river discharge. Although in 1990 the forested land slightly decreased (30%), the irrigation demand was almost double, leading to higher exploitation of underground waters, offering residual water in the rivers' flow and leading eventually to the highest river discharge. Finally, the increase of forested areas in 2018 (34%) and the additional high irrigation demand in 2018 led to the small decrease of river discharge.

Regarding trend analysis, the effect of land cover change on the trend magnitude was evident. Concerning precipitation and river discharge, the trend change points identified were almost identical. Additionally, the trend change points of actual evapotranspiration identified coincide with those of precipitation, verifying the fact that precipitation is a major factor affecting actual evapotranspiration in dry areas, in contrast to wet areas that evapotranspiration is energy-limited (radiation and air temperature) (for example [77–79]). On the contrary, the trend change points of actual evapotranspiration and air temperature were not the same, indicating that actual evapotranspiration is affected in a more complicated way and also by other factors except air temperature as expected, such as land cover and water availability. This was also evident during the trend magnitude analysis of each trend period, where the effect of land cover was noticeable. More specifically, in the case of LC1960, where mean annual actual evapotranspiration was the highest in comparison to the other land cover cases examined, and forested land and pastures (that also include natural grasslands, sclerophyllous vegetation, transitional woodland-shrub, moors and heathland and sparsely vegetated areas) consisted of 70% of the total watershed area, the trend magnitude of each trend period examined was higher. Additionally, highly vegetated watersheds showed smaller tolerance to changes of hydrometeorological factors regarding actual evapotranspiration. On the contrary, the small trend magnitude of river discharge in LC1960 in comparison to LC1990 and LC2018 indicated that in the case of a highly vegetated river basin, the response of the system to changes of hydrometeorological factors regarding river discharge was milder. It is an important finding because land cover of LC1960 could play a relaxing role on the consequences of extreme weather phenomena, either droughts or floods, which will possibly increase in the future.

It should be noted that some uncertainties arise due to the fact that during the present study precipitation and air temperature were considered to be unaffected by land coverage. This is a weakness of the present methodological approach since the current version of the hydrological model MIKE SHE does not provide the option of a two-way dynamically coupled atmospheric-hydrological modeling. The use of an uncoupled system can lead to overprediction of the change in evapotranspiration caused by land cover use changes in comparison to the use of a coupled model results [80].

5. Conclusions

In this study, the physically-based hydrological model MIKE SHE and Copernicus Climate Change Service E-OBS gridded meteorological dataset were used to analyze the effects of anthropogenic land cover changes to the hydrological cycle components of the regional watershed of Spercheios river in central Greece. Three case studies based on the land cover of the years 1960, 1990, and 2018 were investigated.

The analysis of simulation results showed that phenomena like deforestation reduced mean annual actual evapotranspiration while increasing mean annual river discharge. The increase of irrigated agricultural land and irrigation demand also increased discharge as revealed by the results of the case study based on the latest land cover of 2018. Even though irrigation often reduces overland water resources, the exploitation of underground waters can increase river discharge.

Moreover, the climatic variabilities primarily in precipitation and secondarily in temperature influenced annual actual evapotranspiration and annual river discharge. Nevertheless, the response of various watershed areas on land cover changes was shown to be more significant, hiding the effects of climatic variabilities. Land cover changed among the case studies, and thus, locally exceeded the impact of climatic variabilities as indicated by the reduced interannual variabilities of differences in annual actual evapotranspiration. The inhomogeneity of land cover as well as the reduction of vegetated areas were highlighted as the main reasons for this effect.

Remarkably, an in-depth trend analysis unveiled the effect of land cover on increasing the vulnerability on extreme climatic variabilities causing intense hydrometeorological events, either droughts or floods. This means that the resilience of the watershed to extreme weather and climatic phenomena was higher in cases of increased vegetated area, since the response of river discharge in changes of hydrometeorological factors and precipitation was milder in cases of land cover dominated by forested land. This finding highlights the fact that the natural systems under stress mainly due to land cover changes and anthropogenic interventions are likely to have more rapid and acute reactions to climatic variabilities.

Understating the complex interactions among multiple stressors—land degradation and hydrometeorological hazards—can contribute to the development and implementation of successful Integrated Water Resources Management plans. Given the high level of uncertainty of climate change projections and related impacts on water resources, the effects of climatic variabilities on freshwater resources cannot be quantified in a deterministic way; decision-making should be rather based on possible future freshwater hazards and risks. Under this scope, the quantitative assessment of land cover effects presented in this study can be a basis for adaptation and mitigation to climate change and human interventions.

Author Contributions: Conceptualization, A.M., G.V., E.D., A.P., I.P. and P.K.; Formal analysis, A.M. and G.V.; Investigation, A.M., E.D., A.P., I.P. and P.K.; Methodology, A.M., G.V., E.D., A.P., I.P. and P.K.; Resources, A.M.; Supervision, E.D., A.P., I.P. and P.K.; Validation, A.M. and G.V.; Visualization, A.M. and G.V.; Writing—original draft, A.M., G.V., E.D., A.P., I.P. and P.K.; Writing—review & editing, A.M., G.V., E.D., A.P., I.P. and P.K.

Funding: This research received no external funding.

Acknowledgments: We acknowledge the E-OBS dataset from the EU-FP6 project UERRA (http://www.uerra.eu) and the Copernicus Climate Change Service, and the data providers in the ECA&D project (https://www.ecad.eu).

Conflicts of Interest: The authors declare no conflict of interest.

References

1. Kundzewicz, Z.; Krysanova, V.; Benestad, R.; Hov, Ø.; Piniewski, M.; Otto, I. Uncertainty in climate change impacts on water resources. *Environ. Sci. Policy* **2018**, *79*, 1–8. [CrossRef]
2. Dingman, S.L. *Physical Hydrology*, 3rd ed.; Waveland Press, Inc.: Long Grove, IL, USA, 2015; Volume 2.
3. Zeiringer, B.; Seliger, C.; Greimel, F.; Schmutz, S. River hydrology, flow alteration, and environmental flow. In *Riverine Ecosystem Management*; Schmutz, S., Sendzimir, J., Eds.; Aquatic Ecology Series 8; Springer: Cham, Switzerland, 2018; pp. 67–89.
4. Collier, C.G. *Hydrometeorology*; John Wiley & Sons, Ltd.: Chichester, UK, 2016.
5. Narasimhan, T.N. Hydrological Cycle and Water Budgets. *Encycl. Inland Waters* **2009**, 714–720. [CrossRef]
6. Tegen, I.; Fung, I. Contribution to the Atmospheric Mineral Aerosol Load from Land-Surface Modification. *J. Geophys. Res. Atmos.* **1995**, *100*, 18707–18726. [CrossRef]

7. Cisneros, J.B.E.; Oki, T.; Arnell, N.W.; Benito, G.; Cogley, J.G.; Döll, P.; Jiang, T.; Mwakalila, S. Freshwater resources. In *Climate Change 2014: Impacts, Adaptation, and Vulnerability. Part A: Global and Sectoral Aspects. Contribution of Working Group II to the Fifth Assessment Report of the Intergovernmental Panel on Climate Change*; Field, C.B., Barros, V.R., Dokken, D.J., Mach, K.J., Mastrandrea, M.D., Bilir, T.E., Chatterjee, M., Ebi, K.L., Estrada, Y.O., Genova, R.C., et al., Eds.; Cambridge University Press: Cambridge, UK; New York, NY, USA, 2014; pp. 229–269.
8. Döll, P.; Jiménez-Cisneros, B.; Oki, T.; Arnell, N.W.; Benito, G.; Cogley, J.G.; Jiang, T.; Kundzewicz, Z.W.; Mwakalila, S.; Nishijima, A. Integrating risks of climate change into water management. *Hydrol. Sci. J.* **2015**, *60*, 4–13. [CrossRef]
9. Brovkin, V.; Boysen, L.; Arora, V.K.; Boisier, J.P.; Cadule, P.; Chini, L.; Claussen, M.; Friedlingstein, P.; Gayler, V.; Van den hurk, B.J.J.M.; et al. Effect of anthropogenic land-use and land-cover changes on climate and land carbon storage in CMIP5 projections for the twenty-first century. *J. Clim.* **2013**, *26*, 6859–6881. [CrossRef]
10. Poff, N.L.; Allan, J.D.; Bain, M.B.; Karr, J.R.; Prestegaard, K.L.; Richter, B.D.; Sparks, R.E.; Stromberg, J.C. The Natural Flow Regime. *Bioscience* **1997**, *47*, 769–784. [CrossRef]
11. Haddeland, I.; Heinke, J.; Biemans, H.; Eisner, S.; Flörke, M.; Hanasaki, N.; Konzmann, M.; Ludwig, F.; Masaki, Y.; Schewe, J.; et al. Global water resources affected by human interventions and climate change. *Proc. Natl. Acad. Sci. USA* **2014**, *111*, 3251–3256. [CrossRef]
12. Wada, Y.; Van Beek, L.P.H.; Bierkens, M.F.P. Modelling global water stress of the recent past: On the relative importance of trends in water demand and climate variability. *Hydrol. Earth Syst. Sci.* **2011**, *15*, 3785–3808. [CrossRef]
13. Sterling, S.M.; Ducharne, A.; Polcher, J. The impact of global land-cover change on the terrestrial water cycle. *Nat. Clim. Chang.* **2013**, *3*, 385–390. [CrossRef]
14. Piao, S.; Friedlingstein, P.; Ciais, P.; de Noblet-Ducoudre, N.; Labat, D.; Zaehle, S. Changes in climate and land use have a larger direct impact than rising CO2 on global river runoff trends. *Proc. Natl. Acad. Sci. USA* **2007**, *104*, 15242–15247. [CrossRef]
15. Blöschl, G.; Ardoin-Bardin, S.; Bonell, M.; Dorninger, M.; Goodrich, D.; Gutknecht, D.; Matamoros, D.; Merz, B.; Shand, P.; Szolgay, J. At what scales do climate variability and land cover change impact on flooding and low flows? *Hydrol. Proceses* **2007**, *21*, 1241–1247. [CrossRef]
16. Bosmans, J.H.C.; Van Beek, L.P.H.; Sutanudjaja, E.H.; Bierkens, M.F.P. Hydrological impacts of global land cover change and human water use. *Hydrol. Earth Syst. Sci.* **2017**, *21*, 5603–5626. [CrossRef]
17. Guzha, A.C.; Rufino, M.C.; Okoth, S.; Jacobs, S.; Nóbrega, R.L.B. Impacts of land use and land cover change on surface runoff, discharge and low flows: Evidence from East Africa. *J. Hydrol. Reg. Stud.* **2018**, *15*, 49–67. [CrossRef]
18. Benjamin, K. *Land Use Impacts on Water Resources: A Literature Review*; Food and Agriculture Organization of the United Nations-FAO: Rome, Italy, 2000.
19. Peel, M.C. Hydrology: Catchment vegetation and runoff. *Prog. Phys. Geogr.* **2009**, *33*, 837–844. [CrossRef]
20. Dias, L.C.P.; Macedo, M.N.; Costa, M.H.; Coe, M.T.; Neill, C. Effects of land cover change on evapotranspiration and streamflow of small catchments in the Upper Xingu River Basin, Central Brazil. *J. Hydrol. Reg. Stud.* **2015**, *4*, 108–122. [CrossRef]
21. Arancibia, J.L.P. *Impacts of Land Use Change on Dry Season Flows Across the Tropics*; King's College London: London, UK, 2013.
22. Chotpantarat, S.; Boonkaewwan, S. Impacts of land-use changes on watershed discharge and water quality in a large intensive agricultural area in Thailand. *Hydrol. Sci. J.* **2018**, *63*, 1386–1407. [CrossRef]
23. Zhu, C.; Li, Y. Long-Term Hydrological Impacts of Land Use/Land Cover Change From 1984 to 2010 in the Little River Watershed, Tennessee. *Int. Soil Water Conserv. Res.* **2014**, *2*, 11–21. [CrossRef]
24. Liou, Y.A.; Nguyen, A.K.; Li, M.H. Assessing spatiotemporal eco-environmental vulnerability by Landsat data. *Ecol. Indic.* **2017**, *80*, 52–65. [CrossRef]
25. Nguyen, A.K.; Liou, Y.A.; Li, M.H.; Tran, T.A. Zoning eco-environmental vulnerability for environmental management and protection. *Ecol. Indic.* **2016**, *69*, 100–117. [CrossRef]
26. Pikounis, M.; Varanou, E.; Baltas, E.; Dassaklis, A.; Mimikou, M. Application of the SWAT model in the Pinios river basin under different land-use scenarios. *Glob. Int. J.* **2003**, *5*, 71–79.

27. Varlas, G.; Anagnostou, M.N.; Spyrou, C.; Papadopoulos, A.; Kalogiros, J.; Mentzafou, A.; Michaelides, S.; Baltas, E.; Karymbalis, E.; Katsafados, P. A Multi-Platform Hydrometeorological Analysis of the Flash Flood Event of 15 November 2017 in Attica, Greece. *Remote Sens.* **2019**, *11*, 45. [CrossRef]
28. Li, G.; Zhang, F.; Jing, Y.; Liu, Y.; Sun, G. Response of evapotranspiration to changes in land use and land cover and climate in China during 2001–2013. *Sci. Total Environ.* **2017**, *596–597*, 256–265. [CrossRef] [PubMed]
29. Zacharias, I.; Dimitriou, E.; Koussouris, T. Quantifying land-use alterations and associated hydrologic impacts at a wetland area by using remote sensing and modeling techniques. *Environ. Model. Assess.* **2004**, *9*, 23–32. [CrossRef]
30. UERRA, Uncertainties in Ensembles of Regional ReAnalysis. Available online: http://www.uerra.eu (accessed on 1 June 2019).
31. Copernicus Climate Change Service. Available online: https://climate.copernicus.eu/ (accessed on 1 June 2019).
32. HCMR. *Development of an Integrated Management System for Basin, Coastal and Marine Zones. Results of the Annual Evaluation of Ecological Quality of Each Water Body*; Hellenic Centre for Marine Research: Anavyssos, Greece, 2015.
33. Tselika, V. Form and Development of Prehistoric Settlements in Greece: Spatial Planning and Settlement Patterning. Ph.D. Thesis, Aristotle University of Thessaloniki, Thessaloniki, Greece, 2006.
34. Skoulikidis, N.T. The environmental state of rivers in the Balkans—A review within the DPSIR framework. *Sci. Total Environ.* **2009**, *407*, 2501–2516. [CrossRef]
35. Mentzafou, A.; Vamvakaki, C.; Zacharias, I.; Gianni, A.; Dimitriou, E. Climate change impacts on a Mediterranean river and the associated interactions with the adjacent coastal area. *Environ. Earth Sci.* **2017**, *76*, 259. [CrossRef]
36. DHI. *MIKE SHE User Manual*; Danish Hydraulic Institute: Copenhagen, Denmark, 2019.
37. Juston, J.; Seibert, J.; Johansson, P.-O. Temporal sampling strategies and uncertainty in calibrating a conceptual hydrological model for a small boreal catchment. *Hydrol. Process.* **2009**, *23*, 3093–3109. [CrossRef]
38. Vrugt, J.A.; Gupta, H.V.; Dekker, S.C.; Sorooshian, S.; Wagener, T.; Bouten, W. Application of stochastic parameter optimization to the Sacramento Soil Moisture Accounting model. *J. Hydrol.* **2006**, *325*, 288–307. [CrossRef]
39. Hinkle, D.; Wiersma, W.; Jurs, S. *Applied Statistics for the Behavioral Sciences*, 5th ed.; Houghton Mifflin: Boston, MA, USA, 2003.
40. Moriasi, D.N.; Arnold, J.G.; Van Liew, M.W.; Bingner, R.L.; Harmel, R.D.; Veith, T.L. Model evaluation guidelines for systematic quantification of accuracy in watershed simulations. *Trans. ASABE* **2007**, *50*, 885–900. [CrossRef]
41. Mentzafou, A.; Wagner, S.; Dimitriou, E. Historical trends and the long-term changes of the hydrological cycle components in a Mediterranean river basin. *Sci. Total Environ.* **2018**, *636*, 558–568. [CrossRef]
42. National Statistical Service of Greece. *Distribution of the Country's Area by Basic Categories of Land Use. Pre-Census Data (Data for Pre-Census Communal Bulletin of Agriculture and Livestock Census of 19 March 1961)*; National Statistical Service of Greece: Athens, Greece, 1962.
43. Bossard, M.; Feranec, J.; Otahel, J. *CORINE Land Cover Technical Guide: Addendum 2000. Technical Report No 40*; European Environmental Agency: Copenhagen, Denmark, 2000.
44. European Environmental Agency CORINE Land Cover CLC1990, CLC2018 v20b2. Available online: http://land.copernicus.eu/ (accessed on 10 June 2019).
45. Cornes, R.C.; van der Schrier, G.; van den Besselaar, E.J.M.; Jones, P.D. An Ensemble Version of the E-OBS Temperature and Precipitation Data Sets. *J. Geophys. Res. Atmos.* **2018**, *123*, 9391–9409. [CrossRef]
46. Hofstra, N.; Haylock, M.; New, M.; Jones, P.D. Testing E-OBS European high-resolution gridded data set of daily precipitation and surface temperature. *J. Geophys. Res. Atmos.* **2009**, *114*. [CrossRef]
47. Copernicus Climate Change Service (C3S) Ensembles User Guide. Available online: http://surfobs.climate.copernicus.eu/userguidance/use_ensembles.php (accessed on 1 June 2019).
48. Klok, E.J.; Tank, A.M.G.K. Updated and extended European dataset of daily climate observations. *Int. J. Climatol.* **2009**, *29*, 1182–1191. [CrossRef]
49. Van Engelen, A.; Klein Tank, A.; van der Schrier, G.; Klok, L. Towards an operational system for assessing observed changes in climate extremes. In *European Climate Assessment & Dataset (ECA&D), Report 2008*; KNMI: De Bilt, The Netherlands, 2008.

50. Hargreaves, G.H.; Samani, Z.A. Reference Crop Evapotranspiration from Temperature. *Appl. Eng. Agric.* **1985**, *1*, 96–99. [CrossRef]
51. Shuttleworth, W.J. Evaporation. In *Handbook of Hydrology*; Maidment, D.R., Ed.; McGraw-Hill Book Company: New York, NY, USA, 1993; pp. 4.1–4.53.
52. Koutsoyiannis, D.; Xanthopoulos, T. *Engineering Hydrology*, 3rd ed.; National Technical University of Athens: Athens, Greece, 1999.
53. National Statistical Service of Greece. *Results of Agriculture and Livestock Census of March 19th 1961. Volume I, Issue 2: Central Greece and Evoia*; National Statistical Service of Greece: Athens, Greece, 1965.
54. National Statistical Service of Greece. *Census of Agricultural and Livestock Holdings 2010*; National Statistical Service of Greece: Athens, Greece, 2014.
55. Savva, A.P.; Frenken, K. Crop water requirements and irrigation scheduling. In *Irrigation Manual Module 4*; Water Resources Development and Management Office, FAO Sub- Regional Office for East and Southern Africa: Harare, Zimbabwe, 2002; Volume 4.
56. National Statistical Service of Greece. *Agricultural Statistics of Greece. Year 1990*; National Statistical Service of Greece: Athens, Greece, 1994.
57. National Statistical Service of Greece. *Annual Agricultural Stiatistical Survey: 2017*; National Statistical Service of Greece: Athens, Greece, 2018.
58. Sneyers, R. *On the Statistical Analysis of Series of Observations. Technical Note no. 143*; World Meteorological Organization: Geneva, Switzerland, 1990.
59. McGilchrist, C.A.; Woodyer, K.D. Note on a distribution-free CUSUM technique. *Technometrics* **1975**, *17*, 321–325. [CrossRef]
60. Page, E.S. Continuous Inspection Schemes. *Biometrika* **1954**, *41*, 100–115. [CrossRef]
61. Montgomery, D.C. *Introduction to Statistical Quality Control*, 6th ed.; John Wiley & Sons, Inc.: New York, NY, USA, 2009.
62. Mann, H.B. Nonparametric tests against trend. *Econometrica* **1945**, *13*, 245–259. [CrossRef]
63. Kendall, M.G. *Rank Correlation Methods*; Charles Griffin and Company: London, UK, 1975.
64. Gilbert, R.O. *Statistical Methods for Environmental Pollution Monitoring*; Van Nostrand Reinhold Co.: New York, NY, USA, 1987.
65. Sen, P.K. Estimates of the regression coefficient based on Kendall's Tau. *J. Am. Stat.* **1968**, *63*, 1379–1389. [CrossRef]
66. Machiwal, D.; Jha, M. *Hydrologic Time Series Analysis: Theory and Practice*; Springer: Dordrecht, The Netherlands, 2012.
67. Brooks, K.N.; Ffolliott, P.F.; Magner, J.A. *Hydrology and the Management of Watersheds*, 4th ed.; Wiley-Blackwell: Ames, IA, USA, 2012.
68. Papadopoulos, A.; Katsafados, P. Verification of operational weather forecasts from the POSEIDON system across the Eastern Mediterranean. *Nat. Hazards Earth Syst. Sci.* **2009**, *9*, 1299–1306. [CrossRef]
69. Servuk, B. Spatial and temporal inhomogeneity of global precipitation data. In *Global Precipitations and Climate Changes. Proceedings of the NATO Advanced Research Workshop on Global Precipitations and Climate Change, La Londe les Maures, France, 27 September–1 October 1993*; Desbois, M., Desalmand, F., Eds.; Springer: Berlin/Heidelberg, Germany, 1994; Volume 57, pp. 219–230.
70. Mass, C.F.; Ovens, D.; Westrick, K.; Colle, B.A. Does increasing horizontal resolution produce more skillful forecasts? The results of two years of real-time numerical weather prediction over the Pacific Northwest. *Bull. Am. Meteorol. Soc.* **2002**, *83*, 407–430. [CrossRef]
71. Psomiadis, E. Research of Geomorphological and Environmental Changes in the Sperchios' River Basin Utilizing New Technologies. Ph.D. Thesis, Agricultural University of Athens, Athens, Greece, 2010.
72. Thiessen, A.H. Precipitation averages for large areas. *Mon. Weather Rev.* **1911**, *39*, 1082–1084. [CrossRef]
73. Tolika, K.; Maheras, P.; Anagnostopoulou, C. The exceptionally wet year of 2014 over Greece: A statistical and synoptical-atmospheric analysis over the region of Thessaloniki. *Theor. Appl. Climatol.* **2018**, *132*, 809–821. [CrossRef]
74. Papadopoulos, A.; Katsafados, P.; Kallos, G.; Nickovic, S. The weather forecasting system for POSEIDON—An overview. *Glob. Atmos. Ocean Syst.* **2002**, *8*, 219–237. [CrossRef]

75. HCMR. *Development of an Integrated Management System for Basin, Coastal and Marine Zones. Hydrological and Hydrochemical Model of the Study Area*; Hellenic Centre for Marine Research-Institute of Marine Biological Resources and Inland Waters: Anavyssos, Greece, 2015.
76. Ministry of Develpmenet-NTUA-Institute of Geological and Mining Research-Centre for Research and Planning. *Master Plan for Water Resource Management of the Country*; Ministry of Develpmenet: Athens, Greece, 2003.
77. Milly, P.C.D.; Dunne, K.A. Trends in evaporation and surface cooling in the Mississippi River basin. *Geophys. Res. Lett.* **2001**, *28*, 1219–1222. [CrossRef]
78. Gao, G.; Chen, D.; Xu, C.Y.; Simelton, E. Trend of estimated actual evapotranspiration over China during 1960–2002. *J. Geophys. Res. Atmos.* **2007**, *112*. [CrossRef]
79. Yang, K.; Ye, B.; Zhou, D.; Wu, B.; Foken, T.; Qin, J.; Zhou, Z. Response of hydrological cycle to recent climate changes in the Tibetan Plateau. *Clim. Chang.* **2011**, *109*, 517–534. [CrossRef]
80. Overgaard, J.; Butts, M.; Rosbjerg, D. Improved scenario prediction by using coupled hydrological and atmospheric models. In *Quantification and Reduction of Predictive Uncertainty for Sustainable Water Resources Management: Proceedings of Symposium HS2004 at IUGG 2007, Perugia, July 2007. IAHS Publ. 313*; Boegh, E., Kunstmann, H., Wagener, T., Hall, A., Bastidas, L., Franks, S., Gupta, H., Rosbjerg, D., Schaake, J., Eds.; IAHS Press: Wallingford, UK, 2007; pp. 242–248.

Article

A Climatology of Atmospheric Patterns Associated with Red River Valley Blizzards

Aaron Kennedy [1,*], Alexander Trellinger [2], Thomas Grafenauer [3] and Gregory Gust [3]

1 Department of Atmospheric Sciences, University of North Dakota, Grand Forks, ND 58202, USA
2 NOAA/National Weather Service, Sioux Falls, SD 57104, USA; alex.trellinger@noaa.gov
3 NOAA/National Weather Service, Grand Forks, ND 58203, USA; thomas.grafenauer@noaa.gov (T.G.); gregory.gust@noaa.gov (G.G.)
* Correspondence: aaron.kennedy@und.edu; Tel.: +1-701-777-5269

Received: 12 March 2019; Accepted: 1 May 2019; Published: 6 May 2019

Abstract: Stretching along the border of North Dakota and Minnesota, The Red River Valley (RRV) of the North has the highest frequency of reported blizzards within the contiguous United States. Despite the numerous impacts these events have, few systematic studies exist that discuss the meteorological properties of blizzards. As a result, forecasting these events and lesser blowing snow events is an ongoing challenge. This study presents a climatology of atmospheric patterns associated with RRV blizzards for the winter seasons of 1979–1980 and 2017–2018. Patterns were identified using subjective and objective techniques using meteorological fields from the North American Regional Re-analysis (NARR). The RRV experiences, on average, 2.6 events per year. Blizzard frequency is bimodal, with peaks occurring in December and March. The events can largely be typed into four meteorological categories dependent on the forcing that drives the blizzard: Alberta Clippers, Arctic Fronts, Colorado Lows, and Hybrids. The objective classification of these blizzards using a competitive neural network known as the Self-Organizing Map (SOM) demonstrates that gross segregation of the events can be achieved with a small (eight-class) map. This implies that objective analysis techniques can be used to identify these events in weather and climate model output that may aid future forecasting and risk assessment projects.

Keywords: Blizzards; blowing snow; climatology; self-organizing maps; synoptic typing

---

1. Introduction

The United States (US) National Weather Service (NWS) currently defines blizzards as events that have sustained winds or frequent gusts ≥ 35mph (16 m·s^{-1}) and considerable falling and/or blowing snow that reduces visibilities to ≤ $\frac{1}{4}$ mile (400 m) for periods of three hours or longer. These events are recorded within the National Centers for Environmental Information (NCEI) *Storm Data* publication that is reliant on submissions by the Warning Coordination Meteorologist (WCM) at each NWS forecast office. This publication serves as the official archive of storm events for the country [1,2].

Within the contiguous United States (CONUS), reported blizzards are most common over the Northern Great Plains (NGP) including the region centered on North and South Dakota [1,2]. At a county level, the highest frequencies are found along the border of North Dakota and Minnesota, which, topographically, makes up the Red River Valley (RRV) of the North (Figure 1). To some extent, this is impressive considering population-related reporting biases noted for warm-season hazardous weather events such as tornadoes [3–6]. Alternatively, reporting biases could exist by the NWS County Warning Area (CWA), as noted for warm-season hazards such as hail [7] and wind [8].

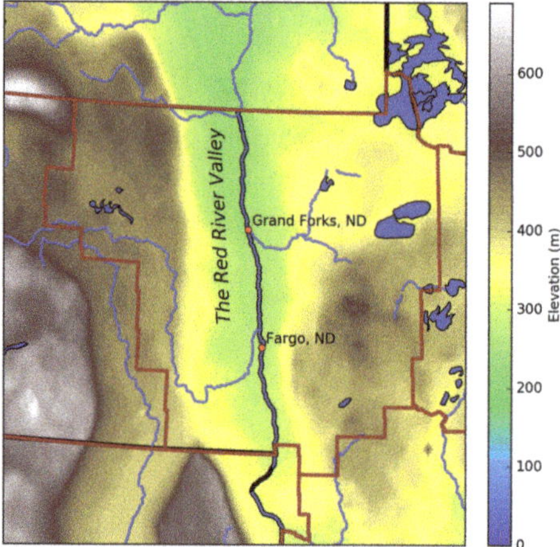

Figure 1. Topography of the Red River Valley (RRV) of the North. Elevation (ASL) is shaded while National Weather Service (NWS) County Warning Areas (CWAs) are denoted by the dark red polygons. Larger water bodies and rivers are highlighted in blue.

Regardless of potential biases in *Storm Data*, the high frequency of blizzards in this region makes physical sense and can be attributed to factors including the topography/land cover, climatology of snow cover, and frequency of high-wind events. A lake plain leftover from the receding Glacial Lake Agassiz 8000 years ago, the shallow Red River of the North flows northward to Lake Winnipeg before eventually emptying into the Hudson Bay [9]. The RRV is largely devoid of trees except within the immediate vicinity of the river and in shelterbelts (tree rows) planted due to agricultural activity. Though the RRV is, on average, only a few hundred meters deep over a 100 km width, there is evidence that winds are enhanced within this region. For example, blowing snow plumes are sometimes seen only within the RRV, typically in regimes with cold-air advection (Figure 2). While there are numerous studies that document topographic influences of valleys on winds in other locations, the authors are unaware of any existing studies for the RRV.

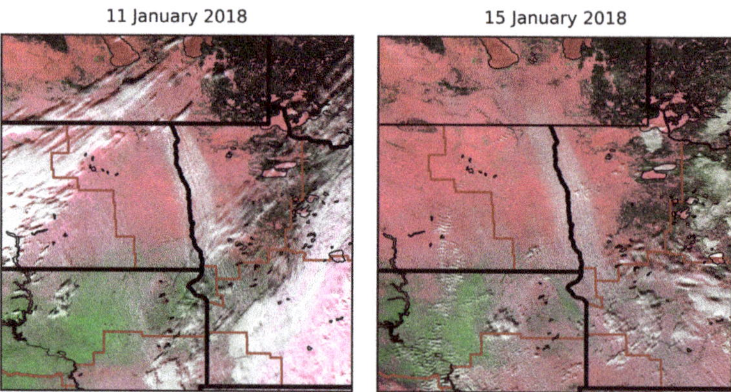

Figure 2. False color imagery (generated from I1-I2-I3-M3-M11 bands) from the Visible Infrared Imaging Radiometer Suite (VIIRS) onboard the Suomi satellite during the daylight (~1:30pm local time) overpass on (**a**) 11 January 2018 and (**b**) 15 January 2018. Snow cover is denoted by pink/red, cloud cover and blowing snow by white, and bare landscape by green (bare ground) or dark (forest) areas. Blowing snow plumes oriented along the RRV are labeled by 'BLSN'.

With a latitude of 45–49° N in the center of the North American continent, the RRV is the coldest non-mountainous region within the CONUS [10]. Though the region only receives, on average, 80–100 cm of snow in a year [11], the cold temperatures facilitate an environment that supports an average snow cover extent >85% during the winter months [12]. Snowfall events responsible for this cover have been tied to several meteorological patterns including extratropical cyclones that form due to lee cyclogenesis such as Colorado Lows and Alberta Clippers [13–17].

As the name implies, Colorado Lows originate due to cyclogenesis near its namesake (the state of Colorado). Historically, these types of systems have been associated with a number of impactful blizzards, including events such as the Children's Blizzard on 12 January 1888 [18]. The strength of these cyclones can advect a significant amount of moisture northward, and, as a result, these systems are responsible for the heaviest (and largest scale) snowfalls in the RRV, southern Manitoba, and western Ontario [16]. The more progressive cousin of these systems include Alberta Clippers that propagate rapidly east-southeast from Canada into the upper-tier of the US [19]. Precipitation for these events typically comes in the form of mesoscale snow bands. Overall, snow totals are lower due to lack of available moisture, but these systems can still produce significant winds capable of reaching blizzard criteria [16,20]. While Colorado Lows and Alberta Clippers are colloquial terms for common North American mid-latitude cyclones, blizzards can also be forced by systems that originate in other areas (e.g., Montana). Historically, these events have been given the moniker 'Hybrids' by the Grand Forks NWS Forecast Office (NWSFO), and, as such, this term is used herein to describe systems that do not conform to stereotypical patterns but have a defined low pressure center. Depending on the event, snowfall can be meso- or synoptic-scale in nature, with high variability for totals.

While blizzards are often thought of as large-scale events associated with the juxtaposition of winds and snowfall associated with mid-latitude cyclones, the RRV also experiences events known as ground blizzards that are frequently driven by strong winds behind Arctic (Cold) Fronts [16,21]. For these cases, winds greater than 4–7 $m{\cdot}s^{-1}$ impart a force on already fallen snow, rolling it on the surface before being bounced and lofted into the atmosphere [22,23]. As demonstrated in Figure 2, these events can often occur under otherwise clear skies and, in some cases, are confined solely to the RRV, providing evidence of the topographic enhancement of winds.

Historically, the Grand Forks NWSFO has subjectively classified blizzards within their CWA (see Figure 1) into the four aforementioned categories (Alberta Clippers, Arctic Fronts, Colorado Lows, and Hybrids), and has maintained a local database of these blizzards from 1974 to present. These events are identical to the reported blizzards in *Storm Data,* although additional meteorological information is sometimes included within the local dataset vs. what is officially provided in *Storm Data*. While *Storm Data* is considered the official dataset for blizzard events, the events are of such importance that the local newspaper (The Grand Forks Herald) has independently kept track of and named impactful events since the winter of 1989–1990.

The purpose of this work is two-fold. First, the climatology of blizzards within the Grand Forks NWSFO CWA will be described for the winters of 1979–1980 and 2017–2018. Besides investigating when and how often blizzards occur, this abbreviated time period will allow for composite patterns to be generated using the North American Regional Re-analysis (NARR) [24]. Given the limitations and known human biases of subjectively defining atmospheric patterns, the second goal of this work is to demonstrate that atmospheric patterns associated with these events can be objectively defined. To do so, a competitive neural network known as a Self-Organizing Map (SOM) [25] will be used.

The efforts of this work will add to the limited body of literature that discuss blizzards in continental regions such as the Northern Great Plains. The climatology and demonstration of an objective technique to classify these patterns described herein will pave the way for future studies that will seek to identify these events in re-analyses, Numerical Weather Prediction (NWP) models, and climate simulations. This will allow for questions to be investigated that range from best forecasting practices for these events to how blizzards may change in a warming climate.

2. Materials and Methods

As noted in the introduction, the climatology of blizzard events in this study comes from the publically available NCEI *Storm Data*. For the purposes of this project, only blizzards contained within the Grand Forks NWSFO CWA (Figure 1) were investigated (see Appendix A). These events are referred to as RRV blizzards due to the majority of the CWA encompassing this topographic feature, although a few counties within this region are on the periphery of the valley. To compare events to NARR data, the time period was limited to the winter seasons (October–April, determined by reported blizzards in Storm Data) of 1979–1980 and 2017–2018. Subjective classifications of event types (Alberta Clippers, Arctic Fronts, Colorado Lows, and Hybrids) were made by Grand Forks NWSFO meteorologists using available observations, model, and re-analysis output.

2.1. Composite Analysis

Composite surface and upper-air patterns were generated using the NARR [24]. Though on a native 32 km horizontal grid, this dataset was averaged to a lower resolution, 16 × 16, and 1.25° (longitude) by 0.94° (latitude) grid centered on the Grand Forks NWSFO CWA. This was done to reduce the computational cost of the SOM and to facilitate future comparisons to other datasets (e.g., weather or climate model output). While a number of re-analyses are now available, NARR was chosen due to the authors' familiarity with this dataset, along with prior studies that demonstrated favorable performance over the region [25–28]. Given the variables and resolution used, it is anticipated that similar results would be found if other current generation re-analyses were used (e.g., ERA-Interim [29]). This assertion may not be valid in regions with more limited surface and upper-air observations where re-analyses are more poorly constrained.

Using storm data, available surface observations, and the NARR, midpoint times were estimated for each blizzard event. Patterns were composited for the four primary patterns using blizzard midpoint times and +/−12-hr before and after these points. For patterns that contained a mid-latitude cyclone, the minimum mean sea level pressure (MSLP) within the domain was identified and tracked over this time.

2.2. Objective Classification Using a SOM

To objectively classify atmospheric patterns, the Self-Organizing Map (SOM) [30] technique was used. A competitive neural network, SOMs are most similar to a K-means clustering algorithm [31]. Unlike K-means clustering, SOMs include a neighborhood function during the training process. The result is a topological (feature) map that allows clusters (nodes) to a) span the data space and b) relate to each other in a two-dimensional matrix. This latter property allows users to be less concerned with the exact number of clusters to choose and instead to focus on clusters that are relevant for their analysis purposes. While this alone makes it a useful algorithm for pattern recognition, SOMs hold other advantages (handling of noise, no a priori assumptions of data, better identification of pattern mixing) over common techniques such as Principal Component Analysis (PCA)/Empirical Orthogonal Functions (EOFs) [32–34]. As a result, SOMs are now commonly used in the fields of meteorology and oceanography. For additional information, the reader is referred to earlier surveys of SOM studies [35,36].

The process of creating a SOM follows the strategy employed in earlier work by the author [37], and the reader is referred to this study for more details on the nuances of SOM creation. To summarize the process, a user must first select data for input then reduce the multi-dimensional meteorological data into input vectors that the SOM performs the clustering on. SOMs are trained in a two-step process that first determines the orientation of the feature map then iterates to a final solution that seeks to minimize the error between the training dataset and the final classification of nodes [31]. These stages require the selection of user parameters such as the map size, training length, learning rate, and the neighborhood radius. After the SOM is created, training samples are compared to each node within the feature map and classified to the node with the minimum Euclidean distance.

Consistent with the generation of composite patterns in the previous section, the spatially averaged 16 × 16, 1.25° (longitude) by 0.94° (latitude) NARR was used to train the SOM. Based on the results of the compositing process, variables that showed significant variability across patterns were used, and these included 500 hPa geopotential heights, MSLP, and surface temperatures. Other combinations of variables were tried, but the inclusion of 500 hPa geopotential heights made the largest difference in the ability of the SOM to segregate patterns. Identical to [37], variables were computed as anomalies from the field mean at each given time step. This allowed the SOM to focus on the gradients in variables, minimizing the issues of biases or variability that vary by season or exist when patterns are compared across multiple datasets (e.g., NARR vs. climate model data) which is useful for future studies. To capture the progression of systems across the domain, each training sample included time steps at the midpoint and +/−12 hrs. With three total variables, a 16 × 16 region, and three times for each case, input vectors used to train the SOM had a length of 2304 elements. All variables were normalized to a common scale to contribute equally to the SOMs. In total, 93 blizzard cases (a subset of 37 winters from 1979–2018 and 2015–2016) were used as input vectors due to the availability of NARR data at the time the SOM was created and as a goal to classify future patterns. Errors (Euclidean distances) for classified patterns in the latter two seasons were similar to the trained data. This suggested that (1) training the SOM with all 100 patterns would not significantly alter the results, and (2) ample variability was captured in the SOM and the methodology is useful for pattern recognition purposes.

A key parameter of the SOM (and other objective classification techniques) is the number of classes chosen. For classifications of atmospheric states, this decision will be dependent on the purpose of the study as well as the number of samples being used to train the SOM. Too few classes can smooth out the details of patterns, while too many will lead to situations where some SOM nodes have no observed patterns classified to them [37]. With a relatively small number of cases (~90–100) and the purpose of comparing the SOM to subjectively classified classes, a rectangular 8-class (4 × 2) is presented. Larger maps were also created, but they did not provide further insight to the results shown herein.

The SOM was generated using SOM_PAK software, which is freely available [30]. Within this package, the 'vfind' program was used; this program randomly initializes a SOM feature map a specified number of times and selects the map that minimizes the lowest quantization error. Following the guidelines of SOM_PAK [30], settings for 'vfind' included a training length that increased and learning rates and neighborhood radii that decreased between the two steps in the training process (Table 1).

Table 1. Self-Organizing Map (SOM) settings used with the SOM_PAK command 'vfind'.

Parameter	Value	Notes
Topology	Rectangular	vs. hexagonal lattice
Neighborhood Function	Bubble	vs. Gaussian
Trials	10	randomly initialized
Training Length (stage 1, stage 2)	93, 93000	# of blizzard patterns
Learning Rate (stage 1, stage 2)	0.05, 0.01	linearly decrease with time
Neighborhood Radius (stage 1, stage 2)	3, 1	# of nodes

3. Results and Discussion

3.1. General Characteristics

During the 39-year period, 100 total blizzards were reported, averaging 2.6 events per year. An annual and seasonal breakdown of these events is provided in Figure 3. RRV Blizzards are highly variable, with seasons varying from 0–10 events (Figure 3a). Record years (10 events) included the infamous 1996–1997 winter that concluded with the catastrophic RRV flood [38,39] and the 2013–2014 winter that did not have significant flooding. On the other end of the spectrum, three seasons (1986–1987, 1990–1991, and 2011–2012) did not have any recorded blizzards. While the source of this variability is beyond the scope of this study, snowfall and cyclone variability in this region have been tied in part to phases of the El Niño Southern Oscillation (ENSO) and the North American Oscillation (NAO) [40,41].

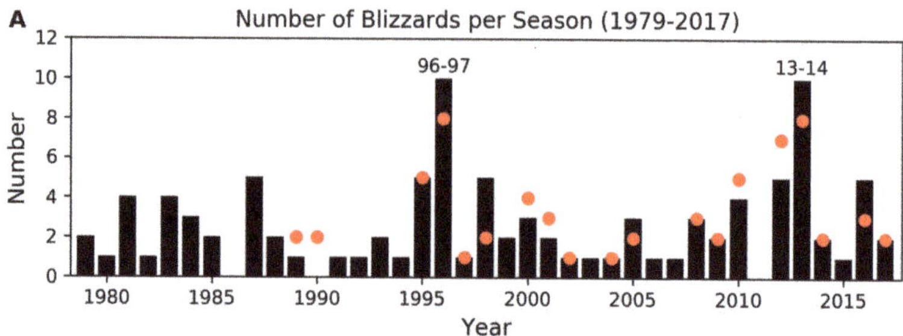

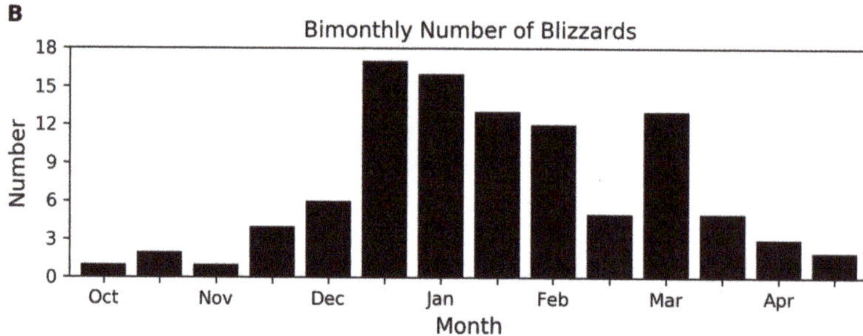

Figure 3. (**a**) Annual and (**b**) bimonthly number of *Storm Data* blizzards for the winter seasons of 1979–1980 and 2017–2018. Named blizzards by the Grand Forks Herald are provided by the red dots in panel (**a**).

Out of curiosity, named blizzards from the Grand Forks Herald newspaper were also compared for annual totals. Over the shorter period (1989–1990 and 2017–2018), the newspaper named 63 (vs. 76) blizzards, and the datasets had a correlation of 0.77. Provided that the distribution area for the paper is smaller than the CWA, these results are expected. While some specific years had more events recorded by the paper vs. *Storm Data*, this is attributed to events that were stronger winter storms but did not meet official blizzard criteria.

Blizzards have been reported from October–April, with the bulk of the events occurring from the 2nd half of December to the 1st half of March (Figure 3b). The most frequent period of occurrence was the 2nd half of December, with a total of 17 blizzards over the 39-year period. A unique aspect of the seasonal cycle is the bimodal distribution with a well-defined lull in late February. This is in agreement with North American cyclone climatologies that indicate a relative minima of cyclones during February [42].

3.2. Composite Analysis

Classifications of the 100 classified blizzards were used to generate composite patterns from the NARR. Of the 100 patterns, two patterns were sufficiently different that they did not fit any of the four categories, and these were omitted from the composite analysis. These included two events driven by southerly winds well ahead of weaker mid-latitude cyclones on 6 March 2014 and 31 December 1996. The remaining composite patterns are now described.

Of the four patterns, Colorado Low blizzards feature the strongest mid-latitude cyclone (Figures 4a–7a) and resemble prior composites of this type [43]. Tracking from Northeast Colorado to North Wisconsin, the composite minimum MSLP decreases from 1002–1000 hPa from 12 h prior to the midpoint of the event. With a storm track south of the RRV, the region is predominately under northerly surface winds that strengthen and shift direction from the East-Northeast to the Northwest as the cyclone progresses eastward. While not shown (and noted earlier), these systems are responsible for the highest snowfall totals, as the region falls within the precipitation shield north of the cyclone track [44]. Aloft, these events are associated with the progression of a well-defined trough that deepens over the region (Figures 5 and 7a). In response to the passage of this trough, strong 500 hPa height falls are found leading up to the event, with the maximum decrease located just northeast of the surface low. In some cases, these troughs are associated with an upper-level closed low, although this definition has been lost to some extent during the compositing process.

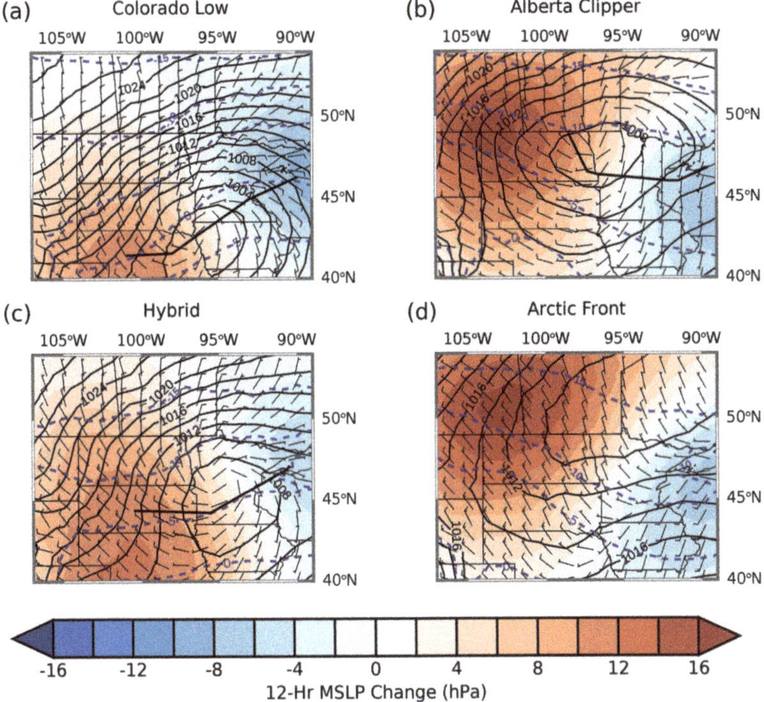

Figure 4. North American Regional Re-analysis (NARR) composite plots of mean sea level pressure (MSLP) (hPa), surface wind barbs (kts), and surface temperatures (°C) 12 hr prior to the midpoint of (**a**) Colorado Low, (**b**) Alberta Clipper, (**c**) Hybrid, and (**d**) Arctic Front blizzards. 12-h MSLP change (midpoint—12 hr prior) is provided by shaded contours while composite mean cyclone tracks are denoted by the thick black lines for select classes.

Blizzards associated with Alberta Clippers also feature a well-defined, albeit weaker (1008–1006 hPa) mid-latitude cyclone (Figures 4b–7b). Consistent with the name and prior composites [19], these systems track east-southeast from southern Canada across the RRV with the cyclone center eventually reaching Northeast Minnesota and North Wisconsin. One should note that the composite cyclone tracks appear to be shifted east compared to the Colorado Low and barely encompass the passage of the low out of Canada. Because these tracks were identified symmetrically around the midpoint of the blizzard conditions, this implies that poor visibility primarily occurs after the passage and development of the surface cyclone (Figure 4b). Aloft, conditions leading up to the event feature stronger west-northwest 500 hPa flow with maximum height falls located over Minnesota, just ahead of a developing short-wave trough (Figure 5b). By the midpoint of the blizzard, this trough has amplified and progressed eastward across the domain with 500 hPa winds shifting to the Northwest.

As noted earlier, the Grand Forks NWS defines Hybrid events as those with characteristics of multiple patterns, and this is also true of the composite patterns (Figures 4c–7c). At the surface, this class manifests itself as a mid-latitude cyclone track that begins farther north (south) of a Colorado Low (Alberta Clipper). The minimum pressure and intensity of the wind field are similar to that of the Alberta Clipper, but it has weaker 12-hr pressure rises/falls (Figure 6c). This latter property can be attributed to the slower progression of Hybrids vs. Alberta Clippers. At 500 hPa, Hybrids are associated with weaker, near-zonal flow 12 h prior to the midpoint of blizzard conditions (Figure 4c). Compared to the Alberta Clippers, the short-wave trough is in a similar position, but the orientation of the flow leads to a more neutral tilt. Unlike the aforementioned pattern, Hybrids feature more deepening of the upper-level low/trough by the midpoint of the blizzard, similar to what is seen for Colorado Lows.

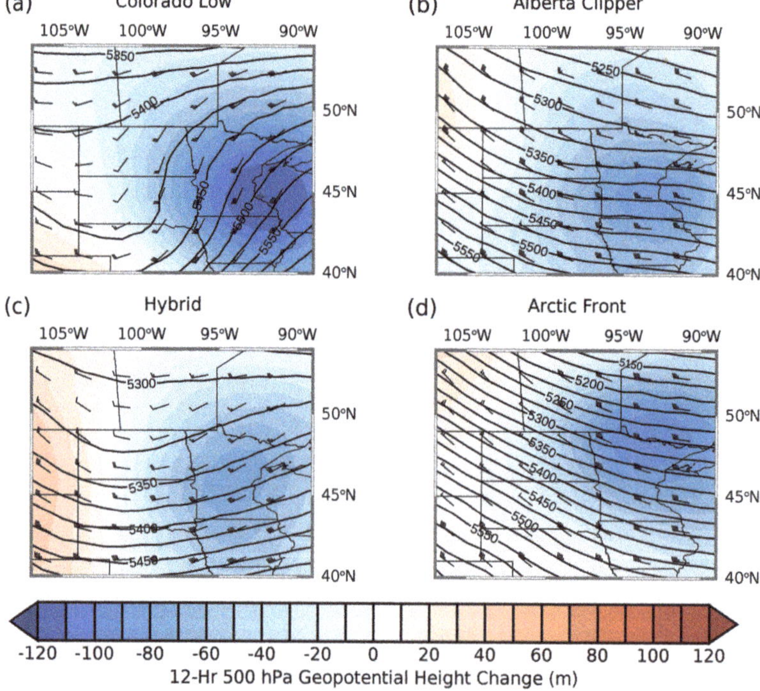

Figure 5. NARR composite plots of 500 hPa Geopotential heights (m) and wind barbs (kts) 12 hr prior to the midpoint of (**a**) Colorado Low, (**b**) Alberta Clipper, (**c**) Hybrid, and (**d**) Arctic Front blizzards. 12-h 500 hPa height change (midpoint—12 hr prior) is provided by shaded contours.

Arctic Fronts are the final, and arguably most unique, composite pattern identified with RRV blizzards (Figures 4d–7d). Unlike the other patterns, no centralized region of low-pressure is seen at the surface. Instead, this pattern features an elongated southwest to northeast oriented surface trough associated with the developing Arctic Front (Figure 4d). As the event progresses, surface pressures rapidly rise and northerly winds strengthen behind the front, as the Arctic High develops to the Northwest, increasing the gradient in MSLP and Cold Air Advection (CAA). Because the surface trough/Arctic Front is often associated with a more distant cyclone, 500 hPa patterns are more dissimilar from the other patterns (Figures 5 and 7d). These events are characterized by strong northwest flow that eventually develops a southwest trough (passing vorticity maxima) in the eastern half of the domain by the mid-point of the event. As a result, the region ends up residing under large (60 m) height rises associated with a strengthening jet stream and implied Negative Vorticity Advection (NVA). Compared to the Alberta Clippers and Hybrid events, 500 hPa winds associated with Arctic Fronts are approximately double in magnitude (80 vs. 40 kts). This leads to a vertical wind profile (not shown) that implies downward transfer of momentum in a regime of subsidence is a key mechanism for reaching blizzard criteria for winds. The presence of CAA, NVA, and subsidence matches many of the checklist items for impactful post-cold frontal winds [21].

Meteorological patterns responsible for blizzard events have preferred periods of occurrence (Figure 8). Early and late season events (October, November, and April) are primarily due to Hybrid and Colorado Lows, with only one (Alberta Clipper) event not fitting these categories. These classes have bimodal distributions with Colorado Lows (Hybrids) peaking in December and March (January and March), respectively. Alberta Clippers occur from December to March, with the majority of the events occurring during January and February, consistent with [19]. Arctic Fronts, commonly responsible for ground blizzards, are more common during the late winter with events between December to March and a maximum in January. As a result of these distributions, January ends up being the most diverse month with relatively constant fractions (0.2–0.3) across the categories (Figure 8b). As noted earlier, the lull in February is consistent with extra-tropical cyclone climatologies, and this is seen in Figure 8 as a reduction of Colorado and Hybrid lows in this month.

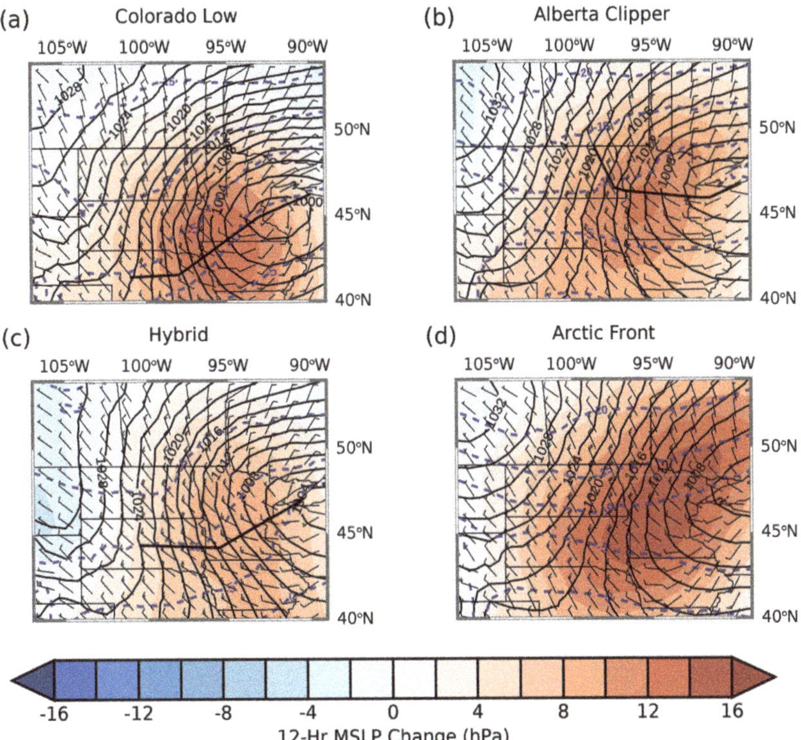

Figure 6. As in Figure 4, except for the midpoint of the blizzard. 12 hr MSLP change (12 hr post—midpoint) is provided by shaded contours.

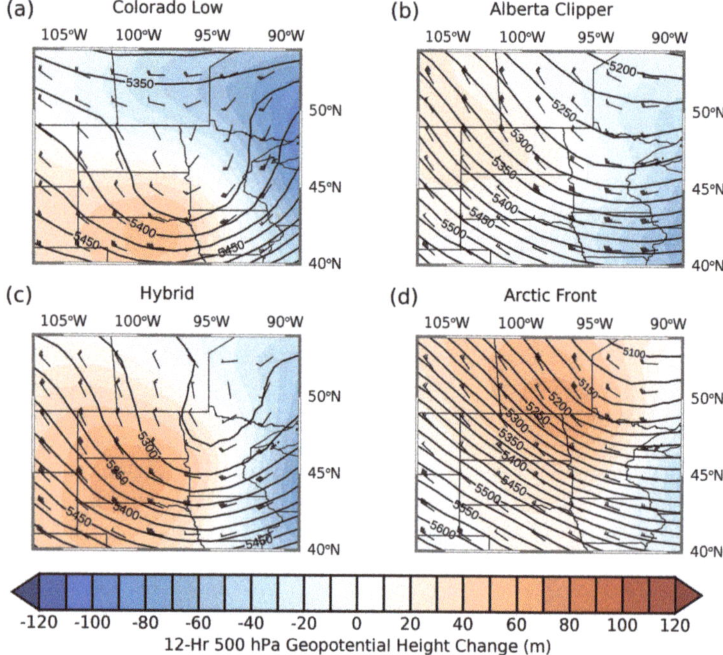

Figure 7. As in Figure 5, except for the midpoint of the blizzard. 12-h 500 hPa height change (12 hr post—midpoint) is provided by shaded contours.

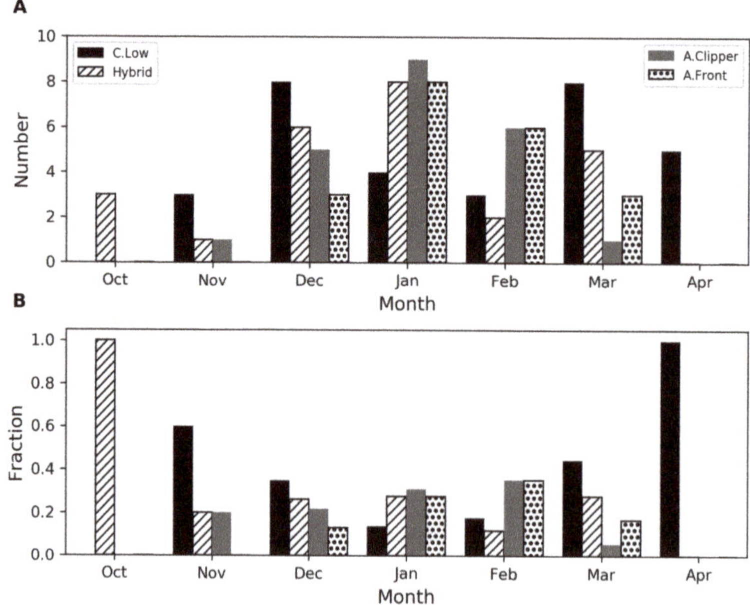

Figure 8. (**a**) Number and (**b**) fraction of monthly blizzards for the winter seasons of 1979–1980 and 2017–2018, separated by type.

3.3. Objective Classification of Patterns Using a SOM

Surface and 500 hPa analyses for the midpoint of blizzard events in the eight-class SOM are shown in Figures 9 and 10. The SOM shows a progression of patterns that shift from cold fronts associated with CAA (nodes 1/5) to deeper surface lows (nodes 3/4/7/8). These patterns have 500 hPa analyses similar to those seen in earlier composites. For example, nodes 1/5 resemble Arctic Front patterns with a strong northwesterly flow aloft, while the rightmost nodes appear as Colorado Lows with either upper-level toughing (nodes 3/4) or a closed low (node 7/8). The progression of systems is also similar to the composites shown earlier (not shown). For example, the rightmost nodes (4/8) progress northeastward like a Colorado Low. Shifting from right-to-left, the mid-latitude cyclones become weaker and have tracks that are displaced northerly, consistent with Hybrid/Alberta Clipper type systems. While the SOM has many positive traits when compared to the composites, it by no means is a perfect reproduction of the subjectively classified classes. For example, Arctic Fronts have pressures that are too low 12-hrs prior to the midpoint of events resulting in weak cyclones vs. the open trough seen in Figure 4. This is undoubtedly a result of the neighborhood function within the SOM smoothing this category with other mid-latitude cyclone nodes. Despite this issue, quantitative comparisons of blizzards events to both the composite patterns, and the eight-class SOM yielded better agreement for the SOM with mean Euclidean distance ~30% lower (315.6 vs. 444.1). Increasing the SOM to a 5 × 3 (or larger) map mitigates this issue and further lowers the mean Euclidean distance; however, this happens at the expense of decreasing the number of blizzards that occur per class (not shown).

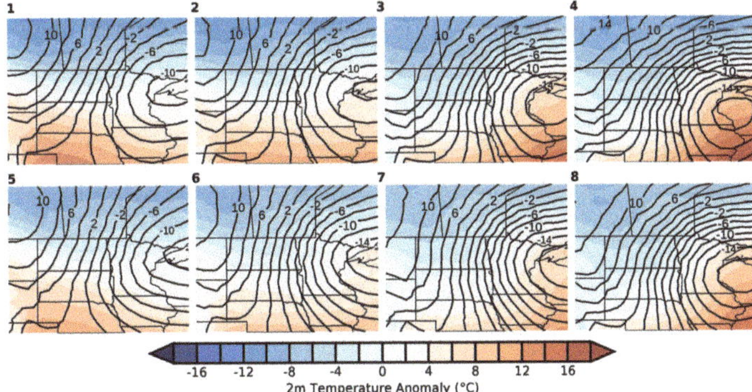

Figure 9. MSLP (hPa, dashed lines) and surface temperature (°C, filled contours) anomalies during the midpoint of blizzards for the eight-class (2 × 4) SOM. Nodes are identified by the external numbers ranging from 1–4 (5–8) for the top (bottom) rows.

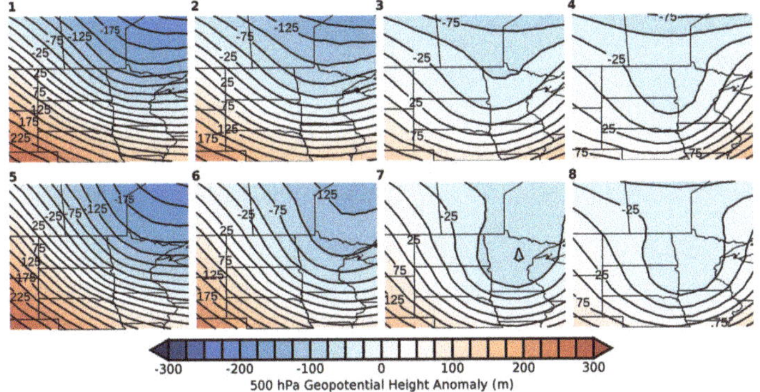

Figure 10. As in Figure 9, except for 500 hPa height anomalies (shaded and dashed contours).

As a final test of the SOM's ability to segregate patterns, the subjectively classified events were categorized into the eight-class SOM (Figure 11). As expected by the meteorological interpretation of the nodes, patterns have distinct areas of occurrence. Colorado Lows (Figure 11a) only occur on the right-hand-side of the SOM, with the majority of the cases occurring within Node 4. Alberta Clippers occur within the left-most six nodes, with most occurring within Nodes 1, 2, 6, and 7. Hybrids, which are subjectively defined as patterns with features of multiple patterns are the only category to occur within every node. That said, the majority of these cases occur within Nodes 3 and 7, in-between Colorado Lows and Alberta Clippers. Finally, Arctic Fronts are primarily on the left-hand-side of the SOM with the majority of the cases occurring in Node 5. From a probability stand-point, categories within the SOM can be arranged in a column fashion, with probability of occurrence shifting from Arctic Fronts (Nodes 1/5), to Alberta Clippers (Nodes 2/6), to Hybrids (Nodes 3/7), to Colorado Lows (Nodes 4/8). By doing this, the time period of occurrence for these categories gives results similar to the results shown in Figure 8 with seasonal occurrence of SOM nodes varying by column (Table 2).

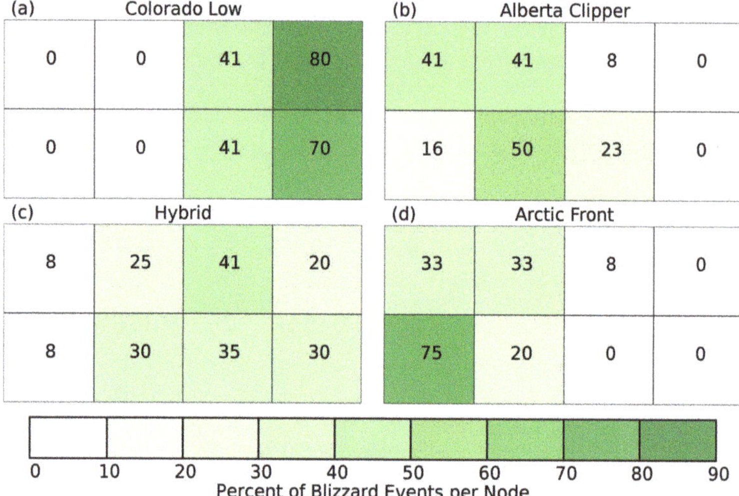

Figure 11. Percent of (**a**) Colorado Low, (**b**) Alberta Clipper, (**c**) Hybrid, and (**d**) Arctic Front blizzards identified within each of the eight SOM nodes.

Table 2. Number (Percentage) of blizzards segregated by month and SOM nodes. Percentages are calculated using monthly totals.

	Nodes 1/5 (Arctic Front)	Nodes 2/6 (Alberta Clipper)	Nodes 3/7 (Hybrid)	Nodes 4/8 (Colorado Low)
October	0 (0)	0 (0)	2 (67)	1 (33)
November	0 (0)	0 (0)	2 (40)	3 (60)
December	2 (9)	7 (30)	9 (39)	5 (22)
January	12 (41)	5 (17)	8 (28)	4 (14)
February	6 (35)	6 (35)	3 (18)	2 (12)
March	4 (22)	4 (22)	4 (22)	6 (33)
April	0 (0)	0 (0)	1 (20)	4 (80)

3.4. Discussion and Future Work

The good agreement between subjectively and objectively identified blizzard patterns provides evidence that the characteristics of these events, including types of patterns and time periods of occurrence, are well understood. The use of a relatively small SOM and inclusion of only several variables to obtain this finding is a positive result that suggests SOMs can be used to investigate a number of outstanding questions regarding blizzards. Some of these activities are now discussed.

Within the realm of weather prediction, a constant struggle is determining whether visibility criteria will be met to justify and verify products such as NWS blizzard warnings. Though blowing snow parametrizations exist [45,46], they are not currently included in operational Numerical NWP models within the US. Instead, local forecasters must use empirical models that determine a probability of blowing snow given conditions such as wind speed, air temperature, and snowpack conditions [47]. The context of how these events fit within the scope of forcing mechanism (type of event) is currently only considered subjectively (e.g., Arctic Fronts are harder to forecast than Colorado Lows). A possible solution is for SOMs to provide real-time identification and classification of forecast atmospheric patterns from deterministic or ensemble NWP systems. While this could be done subjectively, burden is placed on the forecaster to identify patterns within the large range of modeling systems now available. Further, there is always the issue of human bias. A hypothetical system can identify the fractional number of ensemble members with forecasted blizzard patterns. Pattern typing can then inform forecasters on the scope of impacts. For a system such as the Global Ensemble Forecast System (GEFS), doing this subjectively would require the forecaster to individually inspect 21 members, a process that is too arduous for an operational setting.

Prior to the implementation of such a system, the SOM must consider null cases to understand the nuances between patterns that do and do not produce blizzard conditions. In practice, this can be done in two different ways. As presented here, a small SOM developed solely from blizzard events can be used by defining a threshold Euclidean distance that counts a pattern as a hit. A caveat with the presented SOM is distances that are still quite large; this would lead to a higher risk of false alarms. Instead, a larger SOM could be used to reduce the threshold Euclidean distance needed. Alternatively, SOMs can be produced using all available patterns during the winter. To encompass the increased variability in patterns (blizzard and null cases alike), these SOMs must be larger. Instead of using a threshold to define events, observed blizzard events can be mapped to the larger SOM to understand which nodes are associated with these events.

Regardless of the exact methods, an interesting avenue of future work is a retrospective analysis on all historical patterns. This can provide insight into events that may have been missed by the observation system or were too limited in scope to fit within the traditional zone/county verification process at the NWS. Pattern recognition can also be extended farther back in time using datasets such as the 20th Century Re-analysis [48] to yield a long-term climatology of blizzards.

How the frequency and intensity of RRV blizzards may change in a warming climate is also unknown. Previous studies have focused on how precipitation or cyclone frequency may change independently. From the Clausius–Clapeyron relationship, a warmer climate will dictate higher amounts of column water vapor and, thus, precipitation [49]. During the winter, however, there will be a balance between warmer temperatures, column water vapor, and precipitation phase. Overall, a general decline in snow cover has been found for the northern hemisphere, and much of this is due to a significant shortening of the snowy season [50,51]. Despite this trend, the RRV region has seen an increase in snowfall, especially for higher end events with 2+ inches [52,53].

Regarding forcing mechanisms for RRV blizzards, mixed results have been found for extratropical cyclones. While some studies suggest a decrease in Northern Hemisphere wintertime cyclone frequency [54,55], other work suggests the strongest cyclones have intensified or could further intensify in future climate projections [56–59]. Concerning specific patterns identified within the present study, there is a projected decrease (increase) in Alberta Clippers (Colorado Lows) over North America [59]. It is unknown how Hybrid lows or Arctic Fronts may change, and this is an avenue of work into which SOMs can provide insight, as they can provide information on type and frequency of occurrence of patterns.

4. Summary

A climatology of documented blizzard events within Storm Data for the Grand Forks NWSFO CWA for the winter seasons of 1979–1980 and 2017–2018 was presented. The NARR was used to composite and objectively classify patterns. These results are now summarized.

- Over the past 39 years, 100 documented blizzards were reported in *Storm Data*, resulting in an average of 2.6 blizzards per year. This dataset strongly correlates with an unofficial record of societally impactful events named by the Grand Forks Herald, a local newspaper.
- RRV blizzards occur between October and April and have a distinct bimodal distribution of occurrence, with 58% of the events occurring from December 15th to February 15th. After a lull in late February, a separate (weaker) maxima occurs in March.
- The Grand Forks NWSFO has subjectively classified blizzard patterns into four classes: Alberta Clippers, Arctic Fronts, Colorado Lows, and Hybrids. Composite patterns resemble the expected meteorological patterns with variations in the intensity, position, and progressiveness of the mid-latitude cyclone and upper-level trough. Hybrids appear as lows that have tracks in-between the Alberta Clipper and Colorado Low systems.
- Patterns have seasonal variability, with most early/late season blizzards caused by Colorado and Hybrid Lows. Alberta Clippers and Arctic Fronts are more common in the middle of the winter with peak occurrence of these latter patterns in January–February.
- A relatively simple eight-class (4 × 2) SOM can reproduce the general characteristics of the composite patterns. A transition in patterns is seen from Colorado Lows → Hybrids → Alberta Clippers → Arctic Fronts. This results in reasonable separation of subjectively identified events and good agreement in the seasonality of these patterns. This adds confidence to the subjective classification of patterns.

While these results are most relevant to the local populace, the last point has important ramifications for the broader weather and climate communities. Impactful weather events such as blizzards are challenging to forecast/detect over both short and long time-scales due to properties (e.g., visibility) that are not explicitly simulated by weather and climate models. The success of the SOM technique to objectively classify patterns suggests that pattern recognition can be used to address problems such as the predictability of hazardous weather events in NWP ensembles or trends in these events in climate simulations. These subjects are the topics of forthcoming work.

Author Contributions: Conceptualization, A.K. and A.T.; Formal analysis, A.T.; Funding acquisition, A.K.; Investigation, A.T.; Methodology, A.K.; Project administration, A.K.; Supervision, T.G. and G.G.; Visualization, A.T.; Writing—original draft, A.K.; Writing—review and editing, A.K., T.G. and G.G.

Funding: This research was funded by the National Science Foundation Project IIA-1355466 at the University of North Dakota.

Acknowledgments: This paper is dedicated to the late Dave Kellenbenz, a general forecaster at the Grand Forks NWSFO who passed away in 2016 after a courageous battle with Melanoma. Dave maintained the database of blizzards at this office and provided this dataset to the authors prior to his passing that motivated much of this work. The list of named blizzards from the Grand Forks Herald was provided by reporter Tess Williams. NARR data was provided by the NOAA/OAR/ESRL PSD, Boulder, Colorado, USA, from their Web site at http://www.esrl.noaa.gov/psd/.

Conflicts of Interest: The authors declare no conflict of interest.

Appendix A

Table A1. *Storm Data* Blizzards in the Grand Forks NWSFO CWA 1979–1980 and 2017–2018. Italicized events were not included in the typing or SOM analyses. Blizzards named by the Grand Forks Herald but not listed within *Storm Data* are not provided. These events were most likely winter storms with visibility, wind speed, or duration not meeting NWS thresholds.

Year	Month	Day	Midpoint Hour in NARR (UTC)	Type	Grand Forks Herald Name (1989–2018)
2018	1	11	0	Front	Betsy
2017	12	4	21	Colorado	Axl
2017	3	7	6	Hybrid	
2017	1	12	18	Front	Carrie
2016	12	26	15	Colorado	Blitzen
2016	12	7	12	Hybrid	Alvin
2016	11	18	12	Colorado	
2016	2	8	9	Clipper	
2015	1	8	21	Clipper	Beryl
2015	1	3	12	Clipper	Andrew
2014	3	31	21	Colorado	Gigi
2014	3	21	15	Front	
2014	*3*	*6*	*0*	*Ground*	
2014	2	26	21	Front	
2014	2	13	12	Clipper	Fred
2014	1	26	18	Clipper	Era Bell
2014	1	22	12	Front	Dillon
2014	1	16	12	Front	Corene
2014	1	4	6	Clipper	Bubba
2013	12	28	21	Front	Anita
2013	3	18	9	Hybrid	Fiona
2013	2	18	21	Hybrid	Dolley
2013	2	11	3	Colorado	Cooper
2013	1	19	18	Front	Beth
2013	1	12	3	Colorado	Aaron
2011	3	12	6	Clipper	Estra
2011	1	1	9	Colorado	Dave
2010	12	30	21	Colorado	Casey
2010	10	27	9	Hybrid	Adeline
2010	1	25	18	Clipper	Brett
2009	12	26	3	Colorado	Alvin
2009	3	10	21	Colorado	Coyote
2009	1	12	15	Clipper	Barack
2008	12	14	15	Colorado	
2008	2	9	18	Front	
2007	3	3	0	Hybrid	
2006	1	24	15	Front	
2005	11	16	3	Clipper	York
2005	10	6	3	Hybrid	Zach
2005	1	22	6	Clipper	Ann
2004	2	11	18	Clipper	
2003	2	11	18	Front	Arlys
2001	12	23	0	Colorado	Bonnie
2001	10	25	0	Hybrid	Al
2001	2	25	12	Colorado	Dale
2000	12	21	3	Clipper	Carol
2000	12	16	15	Hybrid	Bill
2000	3	9	3	Colorado	
1999	12	19	18	Clipper	
1999	4	1	18	Colorado	
1999	3	17	18	Hybrid	
1999	2	12	12	Front	
1998	12	18	21	Clipper	

Table A1. *Cont.*

Year	Month	Day	Midpoint Hour in NARR (UTC)	Type	Grand Forks Herald Name (1989–2018)
1998	11	10	21	Colorado	Alex
1998	3	13	18	Front	Aurora
1997	4	6	12	Colorado	Hannah
1997	3	4	9	Colorado	Gust
1997	1	22	12	Hybrid	Franzi
1997	1	15	21	Front	Elmo
1997	1	10	9	Clipper	Doris
1997	1	5	9	Colorado	
1996	*12*	*31*	*21*	*Valley*	
1996	12	21	12	Front	Christopher
1996	12	18	0	Clipper	Betty
1996	11	17	9	Colorado	Andy
1996	3	25	0	Colorado	Erin
1996	2	27	21	Hybrid	Darrel
1996	2	10	21	Clipper	
1996	1	18	12	Hybrid	Bruno
1995	12	9	0	Hybrid	Anna
1995	2	10	6	Clipper	
1994	4	26	15	Colorado	
1993	12	22	0	Clipper	
1992	12	25	6	Front	
1991	12	14	3	Hybrid	Dagmar
1990	1	11	12	Clipper	Arnie
1989	2	1	6	Front	
1989	1	7	21	Hybrid	
1988	3	12	3	Colorado	
1988	2	14	15	Clipper	
1988	1	24	21	Hybrid	
1988	1	12	15	Hybrid	
1987	12	31	3	Colorado	
1986	4	15	3	Colorado	
1985	11	19	3	Hybrid	
1985	3	4	6	Colorado	
1985	1	25	0	Front	
1984	12	16	18	Colorado	
1984	3	10	15	Front	
1984	2	5	6	Front	
1983	12	25	0	Hybrid	
1983	12	15	9	Hybrid	
1983	3	8	21	Colorado	
1982	4	3	12	Colorado	
1982	3	8	15	Hybrid	
1982	1	23	15	Colorado	
1982	1	10	18	Hybrid	
1981	2	1	12	Colorado	
1980	1	11	15	Hybrid	
1980	1	7	3	Hybrid	

References

1. Schwartz, R.M.; Schmidlin, T.W. Climatology of Blizzards in the Conterminous United States, 1959–2000. *J. Clim.* **2002**, *15*, 1765–1772. [CrossRef]
2. Coleman, J.S.; Schwartz, R.M. An Updated Blizzard Climatology of the Contiguous United States (1959–2014): An Examination of Spatiotemporal Trends. *J. Appl. Meteorol. Climatol.* **2017**, *56*, 173–187. [CrossRef]
3. Doswell, C.A.; Moller, A.R.; Brooks, H.E. Storm spotting and public awareness since the first tornado forecasts of 1948. *Weather Forecast.* **1999**, *14*, 544–557. [CrossRef]
4. Ray, P.S.; Bieringer, P.; Niu, X.; Whissel, B. An improved estimate of tornado occurrence in the central plains of the United States. *Mon. Weather Rev.* **2003**, *131*, 1026–1031. [CrossRef]

5. Anderson, C.J.; Wikle, C.K.; Zhou, Q. Population influences on tornado reports in the United States. *Weather Forecast.* **2007**, *22*, 571–579. [CrossRef]
6. Elsner, J.B.; Michaels, L.E.; Scheitlin, K.N.; Elsner, I.J. The Decreasing Population Bias in Tornado Reports across the Central Plains. *Weather Clim. Soc.* **2013**, *5*, 221–232. [CrossRef]
7. Allen, J.T.; Tippett, M.K. The characteristics of United States hail reports: 1955–2014. *Electron. J. Sev. Storms Meteorol.* **2015**, *10*, 1–31.
8. Weiss, S.J.; Hart, J.A.; Janish, P.R. An examination of severe thunderstorm wind report climatology: 1970–1999. In Proceedings of the 21st Conference Severe Local Storms, San Antonio, TX, USA, 11–16 August 2002; pp. 446–449.
9. Teller, J.T. Proglacial lakes and the southern margin of the Laurentide Ice Sheet. In *North America and Adjacent Oceans during the Last Deglaciation*; Ruddiman, W.F., Wright, H.E., Jr., Eds.; Geological Society of America: Boulder, CO, USA, 1987; pp. 39–69.
10. Climate Data Online (CDO). Available online: https://www.ncdc.noaa.gov/cdo-web/ (accessed on 10 December 2018).
11. Kluver, D.; Mote, T.; Leathers, D.; Henderson, G.R.; Chan, W.; Robinson, D.A. Creation and Validation of a Comprehensive 1° by 1° Daily Gridded North American Dataset for 1900–2009: Snowfall. *J. Atmos. Ocean. Technol.* **2016**, *33*, 857–871. [CrossRef]
12. Estilow, T.W.; Young, A.H.; Robinson, D.A. A long-term Northern Hemisphere snow cover extent data record for climate studies and monitoring. *Earth Syst. Sci. Data* **2015**, *7*, 137–142. [CrossRef]
13. Newton, C.W. Mechanisms of circulation change in a lee cyclogenesis. *J. Meteorol.* **1956**, *13*, 528–539. [CrossRef]
14. Reitan, C.H. Frequencies of cyclones and cyclogenesis for North America, 1951–1970. *Mon. Weather Rev.* **1974**, *102*, 861–868. [CrossRef]
15. Zishka, K.M.; Smith, P.J. The climatology of cyclones and anticyclones over North America and surrounding ocean environs for January and July, 1950–77. *Mon. Weather Rev.* **1980**, *108*, 387–401. [CrossRef]
16. Stewart, R.E.; Bachand, D.; Dunkley, R.R.; Giles, A.C.; Lawson, B.; Legal, L.; Miller, S.T.; Murphy, B.P.; Parker, M.N.; Paruk, B.J.; et al. Winter storms over Canada. *Atmos.-Ocean* **1995**, *33*, 223–247. [CrossRef]
17. Hoskins, B.J.; Hodges, K.I. New perspectives on the Northern Hemisphere winter storm tracks. *J. Atmos. Sci.* **2002**, *59*, 1041–1061. [CrossRef]
18. Laskin, D. *The Children's Blizzard*, 3rd ed.; HarperCollins: New York, NY, USA, 2005; p. 336.
19. Thomas, B.C.; Martin, J.E. A Synoptic Climatology and Composite Analysis of the Alberta Clipper. *Weather Forecast.* **2007**, *22*, 315–333. [CrossRef]
20. Schultz, D.M.; Doswell, C.A. Analyzing and Forecasting Rocky Mountain Lee Cyclogenesis Often Associated with Strong Winds. *Weather Forecast.* **2000**, *15*, 152–173. [CrossRef]
21. Kapela, A.F.; Leftwich, P.W.; Van Ess, R. Forecasting the impacts of strong wintertime post-cold front winds in the northern plains. *Weather Forecast.* **1995**, *10*, 229–244. [CrossRef]
22. Mellor, M. *Blowing Snow*; US CRREL Monogr: Hanover, NH, USA, 1965; p. 79.
23. Li, L.; Pomeroy, J.W. Probability of occurrence of blowing snow. *J. Geophys. Res.* **1997**, *102*, 21955–21964. [CrossRef]
24. Mesinger, F.; DiMego, G.; Kalnay, E.; Mitchell, K.; Shafran, P.C.; Ebisuzaki, W.; Jović, D.; Woollen, J.; Rogers, E.; Berbery, E.H.; et al. North American regional reanalysis. *Bull. Am. Meteorol. Soc.* **2006**, *87*, 343–360. [CrossRef]
25. Pielke, R., Sr.; Nielsen-Gammon, J.; Davey, C.; Angel, J.; Bliss, O.; Doesken, N.; Cai, M.; Fall, S.; Niyogi, D.; Gallo, K.; et al. Documentation of uncertainties and biases associated with surface temperature measurement sites and for climate change assessment. *Bull. Am. Meteorol. Soc.* **2007**, *88*, 913–928. [CrossRef]
26. Choi, W.; Kim, S.J.; Rasmussen, P.F.; Moore, A.R. Use of the North American Regional Reanalysis for hydrological modelling in Manitoba. *Can. Water Resour. J.* **2009**, *34*, 17–36. [CrossRef]
27. Kim, S.J. Evaluation of surface climate data from the North American Regional Reanalysis for Hydrological Applications in Central Canada. Ph.D. Thesis, University of Manitoba, Winnipeg, Canada, 2012.
28. King, A.T.; Kennedy, A.D. North American Supercell Environments in Atmospheric Reanalyses and RUC-2. *J. Appl. Meteorol. Climatol.* **2019**, *58*, 71–92. [CrossRef]

29. Dee, D.P.; Uppala, S.M.; Simmons, A.J.; Berrisford, P.; Poli, P.; Kobayashi, S.; Andrae, U.; Balmaseda, M.A.; Balsamo, G.; Bauer, P.; et al. The ERA-Interim reanalysis: Configuration and performance of the data assimilation system. *Q. J. R. Meteorol. Soc.* **2011**, *137*, 553–597. [CrossRef]
30. Kohonen, T.; Hynninen, J.; Kangas, J.; Laaksonen, J. SOM_PAK: The self-organizing map program package. *Espoo Helsinki Univ. Technol. Lab. Comput. Inf. Sci.* **1996**, *1*, 39–40.
31. MacQueen, J. Others some methods for classification and analysis of multivariate observations. In *Proceedings of the Fifth Berkeley Symposium on Mathematical Statistics and Probability*; University of California Press: Oakland, CA, USA, 1967; Volume 1, pp. 281–297.
32. Hewitson, B.C.; Crane, R.G. Self-organizing maps: Applications to synoptic climatology. *Clim. Res.* **2002**, *22*, 13–26. [CrossRef]
33. Liu, Y.; Weisberg, R.H.; Mooers, C.N.K. Performance evaluation of the self-organizing map for feature extraction. *J. Geophys. Res.* **2006**, *111*, C05018. [CrossRef]
34. Reusch, D.B.; Alley, R.B.; Hewitson, B.C. North Atlantic climate variability from a self-organizing map perspective. *J. Geophys. Res.* **2007**, *112*, DO2104. [CrossRef]
35. Sheridan, S.C.; Lee, C.C. The self-organizing map in synoptic climatological research. *Prog. Phys. Geogr.* **2011**, *35*, 109–119. [CrossRef]
36. Liu, Y.; Weisberg, R.H. A Review of Self-Organizing Map Applications in Meteorology and Oceanography. In *Self-Organizing Maps: Applications and Novel Algorithm Design*; Mwasiagi, J.I., Ed.; InTech: Rijeka, Croatia, 2011; pp. 253–272.
37. Kennedy, A.D.; Dong, X.; Xi, B. Cloud fraction at the ARM SGP site: reducing uncertainty with self-organizing maps. *Theor. Appl. Climatol.* **2016**, *124*, 43–54. [CrossRef]
38. Macek-Rowland, K.M. *1997 Floods in the Red River of the North and Missouri River Basins in North Dakota and Western Minnesota: U.S. Geological Survey Open-File Report*; USGS: Reston, VA, USA, 1997; pp. 97–575.
39. Rannie, W. The 1997 flood event in the Red River basin: Causes, assessment and damages. *Can. Water Resour. J.* **2016**, *41*, 45–55. [CrossRef]
40. Eichler, T.; Higgins, W. Climatology and ENSO-Related Variability of North American Extratropical Cyclone Activity. *J. Clim.* **2006**, *19*, 2076–2093. [CrossRef]
41. Seager, R.; Kushnir, Y.; Nakamura, J.; Ting, M.; Naik, N. Northern Hemisphere winter snow anomalies; ENSO, NAO and the winter of 2009/10. *J. Geophys. Res.* **2010**, *37*, L14703. [CrossRef]
42. Whittaker, L.M.; Horn, L.H. Geographical and seasonal distribution of North American cyclogenesis, 1958–1977. *Mon. Weather Rev.* **1981**, *109*, 2312–2322. [CrossRef]
43. Achtor, T.H.; Horn, L.H. Spring Season Colorado Cyclones. Part I: Use of Composites to Relate Upper and Lower Tropospheric Wind Fields. *J. Clim. Appl. Meteorol.* **1986**, *25*, 732–743. [CrossRef]
44. Schultz, D.M.; Bosart, L.F.; Colle, B.A.; Davies, H.C.; Dearden, C.; Keyser, D.; Martius, O.; Roebber, P.J.; Steenburgh, W.J.; Volkert, H.; et al. Extratropical Cyclones: A Century of Research on Meteorology's Centerpiece. *Meteorol. Monogr.* **2018**, *59*, 16.1–16.56. [CrossRef]
45. Déry, S.J.; Yau, M.K. Simulation of blowing snow in the Canadian arctic using a double-moment model. *Boundary-Layer Meteorol.* **2001**, *99*, 297–316. [CrossRef]
46. Yang, J.; Yau, M.K. A new triple-moment blowing snow model. *Boundary Layer Meteorol.* **2008**, *126*, 137–155. [CrossRef]
47. Baggaley, D.G.; Hanesiak, J.M. An empirical blowing snow forecast technique for the Canadian Arctic and Prairie Provinces. *Weather Forecast.* **2005**, *20*, 51–62. [CrossRef]
48. Compo, G.P.; Whitaker, J.S.; Sardeshmukh, P.D.; Matsui, N.; Allan, R.J.; Yin, X.; Gleason, B.E.; Vose, R.S.; Rutledge, G.; Bessemoulin, P. The twentieth century reanalysis project. *Q. J. R. Meteorol. Soc.* **2011**, *137*, 1–28. [CrossRef]
49. Soden, B.J.; Held, I.M. An assessment of climate feedbacks in coupled ocean–atmosphere models. *J. Clim.* **2006**, *19*, 3354–3360. [CrossRef]
50. Rupp, D.E.; Abatzoglou, J.T.; Hegewisch, K.C.; Mote, P.W. Evaluation of CMIP5 20th century climate simulations for the Pacific Northwest USA. *J. Geophys. Res. Atmos.* **2013**, *118*, 10,884–10,906. [CrossRef]
51. Brown, R.D.; Robinson, D.A. Northern Hemisphere spring snow cover variability and change over 1922–2010 including an assessment of uncertainty. *Cryosphere* **2011**, *5*, 219–229. [CrossRef]
52. Kunkel, K.E.; Palecki, M.A.; Ensor, L.; Easterling, D.; Hubbard, K.G.; Robinson, D.; Redmond, K. Trends in twentieth-century U.S. extreme snowfall seasons. *J. Clim.* **2009**, *22*, 6204–6216. [CrossRef]

53. Kluver, D.; Leathers, D. Regionalization of snowfall frequency and trends over the contiguous United States. *Int. J. Climatol.* **2015**, *35*, 4348–4358. [CrossRef]
54. Lambert, S.J.; Fyfe, J.C. Changes in winter cyclone frequencies and strengths simulated in enhanced greenhouse warming experiments: Results from the models participating in the IPCC diagnostic exercise. *Clim. Dyn.* **2006**, *26*, 713–728. [CrossRef]
55. Catto, J.L.; Shaffrey, L.C.; Hodges, K.I. Northern Hemisphere extratropical cyclones in a warming climate in the HiGEM high-resolution climate model. *J. Clim.* **2011**, *24*, 5336–5352. [CrossRef]
56. McCabe, G.; Clark, M.; Serreze, M. Trends in Northern Hemisphere surface cyclone frequency and intensity. *J. Clim.* **2001**, *14*, 2763–2768. [CrossRef]
57. Mizuta, R.; Matsueda, M.; Endo, H.; Yukimoto, S. Future change in extratropical cyclones associated with change in the upper troposphere. *J. Clim.* **2011**, *24*, 6456–6470. [CrossRef]
58. Long, Z.; Perrir, W.; Gyakum, J.; Laprisee, R.; Caya, D. Scenario changes in the climatology of winter midlatitude cyclone activity over eastern North America and the northwest Atlantic. *J. Geophys. Res.* **2009**, *114*, D12111. [CrossRef]
59. Eichler, T.P.; Gaggini, N.; Pan, Z. Impacts of global warming on Northern Hemisphere winter storm tracks in the CMIP5 model suite. *J. Geophys. Res. Atmos.* **2013**, *118*, 3919–3932. [CrossRef]

Article

GPS Precipitable Water Vapor Estimations over Costa Rica: A Comparison against Atmospheric Sounding and Moderate Resolution Imaging Spectrometer (MODIS)

Polleth Campos-Arias [1], Germain Esquivel-Hernández [1,*], José Francisco Valverde-Calderón [2], Stephanie Rodríguez-Rosales [2], Jorge Moya-Zamora [2], Ricardo Sánchez-Murillo [1] and Jan Boll [3]

1 Stable Isotope Research Group, School of Chemistry, Universidad Nacional, Heredia 86-3000, Costa Rica; poll3th.ca14@gmail.com (P.C.-A.); ricardo.sanchez.murillo@una.cr (R.S.-M.)

2 School of Topography, Surveying, and Geodesy, Universidad Nacional, Heredia 86-3000, Costa Rica; jose.valverde.calderon@una.cr (J.F.V.-C.); steph_0291@hotmail.com (S.R.-R.); jorge.moya.zamora@una.cr (J.M.-Z.)

3 Department of Civil and Environmental Engineering, Washington State University, Pullman, WA 99164, USA; j.boll@wsu.edu

* Correspondence: germain.esquivel.hernandez@una.cr; Tel.: +506-2277-3484

Received: 20 February 2019; Accepted: 5 April 2019; Published: 3 May 2019

Abstract: The quantification of water vapor in tropical regions like Central America is necessary to estimate the influence of climate change on its distribution and the formation of precipitation. This work reports daily estimations of precipitable water vapor (PWV) using Global Positioning System (GPS) delay data over the Pacific region of Costa Rica during 2017. The GPS PWV measurements were compared against atmospheric sounding and Moderate Resolution Imaging Spectrometer (MODIS) data. When GPS PWV was calculated, relatively small biases between the mean atmospheric temperatures (T_m) from atmospheric sounding and the Bevis equation were found. The seasonal PWV fluctuations were controlled by two of the main circulation processes in Central America: the northeast trade winds and the latitudinal migration of the Intertropical Convergence Zone (ITCZ). No significant statistical differences were found for MODIS Terra during the dry season with respect GPS-based calculations ($p > 0.05$). A multiple linear regression model constructed based on surface meteorological variables can predict the GPS-based measurements with an average relative bias of -0.02 ± 0.19 mm/day ($R^2 = 0.597$). These first results are promising for incorporating GPS-based meteorological applications in Central America where the prevailing climatic conditions offer a unique scenario to study the influence of maritime moisture inputs on the seasonal water vapor distribution.

Keywords: atmospheric sounding; Costa Rica; GPS; MODIS; precipitable water vapor

1. Introduction

Although it constitutes only 0.001% of the planet's water resources, water vapor plays an important role in atmospheric processes as it is one of the major radiative gases and a dynamic element in the atmosphere. Water vapor is a useful parameter to forecast severe weather conditions and precipitation formation and is also a key factor for studying the global water cycle, changing climatic conditions, and earth-atmosphere energy exchange [1–3]. Overall, water vapor is essential for the development of disturbed weather and influences the planetary radiative balance. In the lower atmosphere, it controls the heat exchange during the precipitation formation and the thermal structure of the troposphere, and it is the main source for precipitation in all weather systems [3,4]. Therefore, accurate estimates of atmospheric water vapor content are needed to improve the predictability of rainfall and the understanding of and feedback in climate related processes [5,6].

A quantifiable parameter useful for studying water vapor is the precipitable (or integrated) water vapor (PWV). Precipitable water vapor mainly comprises tropospheric water vapor and the less abundant stratospheric water and can be used to analyze water vapor variability and its contributions to climate change [6]. The classical approach to gather information about PWV is using atmospheric sounding based on radiosonde profiles [7]. However, due to high costs, radiosonde networks lack spatial and temporal resolutions and, thus, provide limited information to carry out detailed studies of weather and climate. For example, radiosondes are usually launched 1–2 times per day in monitoring stations spaced several hundred kilometers from each other. In recent years, the fast development of ground-based GPS networks allows a new source of water vapor information. As atmospheric water changes the atmospheric refractivity, satellite-receiver path delays provide a unique information on the total water vapor within the troposphere and stratosphere. Therefore, GPS has become a standard technique for measuring PWV with some noticeable advantages over radiosondes. For instance, GPS can be used in all weather conditions and has low operation costs, allowing for a high temporal resolution with numerous records throughout the daytime and nighttime [8–10]. In Costa Rica, there are 14 Global Navigation Satellite System (GNSS) stations in operation which are associated with the Sistema de Referencia Geocéntrico para las Américas (SIRGAS) network. Eight of these GNSS stations are officially administrated by the National Institute of Geography. Although there are other GPS stations operating in the country, access to these GPS data is rather limited.

Satellite remote sensing is also a feasible method to derive the PWV distribution. The Moderate Resolution Imaging Spectroradiometer (MODIS) installed at the Terra and Aqua satellites offers spatial and temporal PWV estimations [11,12]. Despite the high spatial coverage and resolution that these satellite-based PWV products offer, there are several sources of errors in water vapor column retrievals from these remote sensing platforms. These errors are mainly linked to an uncertainty in the spectral reflectance of the surface, an uncertainty in the sensor calibration, an uncertainty in the atmospheric temperature and moisture profile, and an uncertainty in the amount of haze [13,14]. Moreover, there are two other additional limitations related to a polar orbiting satellite like MODIS: i) most areas are sampled only once per day, depending on the latitude and the configuration of the instrument, and ii) the measurements are mainly restricted to cloud-free areas (especially during daytime) as clouds are opaque in the visible and NIR spectrum [15]. Unlike satellite-based water vapor estimations, the presence of clouds and precipitation does not affect GPS observations because the liquid water contribution to the refractivity is normally small, especially outside of clouds [16]. In order to assess the performance of satellite measurements, their PWV estimates have been evaluated against other conventional techniques (e.g., GPS PWV measurements) in several regions, for instance in China, in Spain, and in Tibet [6,9,12]. Nevertheless, limited knowledge exists for the Central American Isthmus regarding the application of remote sensing for PWV measurements and how well GPS delay data compare to classical water vapor measurements made by atmospheric sounding in complex tropical mountainous regions like those found in Costa Rica.

In this study, the objectives were i) to evaluate the GPS-based estimates of PWV against PWV based on radiosonde measurements and on the MODIS satellite radiometer, ii) to estimate the influence of the main circulation patterns in Costa Rica on the PWV variability using GPS-based estimations, and iii) to identify major meteorological variables controlling PWV seasonal variations. We selected two GPS stations located in the Pacific region of Costa Rica to calculate the mean daily PWV estimates during 2017. These GPS-based estimations were then compared to PWV measurements made using radiosondes at the only atmospheric sounding site in operation in Costa Rica, located in the Central Valley of the country. We further compared data from the MODIS satellite radiometer against the GPS and radiosonde estimations over the Central Valley of Costa Rica. GPS PWV estimates were also analyzed in combination with surface meteorological data and the Hybrid Single Particle Lagrangian Integrated Trajectory (HYSPLIT) model. We expect that this work will contribute to highlighting the opportunity of incorporating GPS-based meteorological applications in Central America, which can be useful to study the influence of maritime moisture inputs from the Caribbean Sea and Pacific Ocean on the seasonal water vapor distribution.

2. Materials and Methods

2.1. Climatic Characteristics of Costa Rica

Costa Rica is located in the tropics between 8°–11°N latitude and 82°–86°W longitude (Figure 1). The climate of Costa Rica is influenced by four regional air circulation types: NE trade winds, the latitudinal migration of the Intertropical Convergence Zone (ITCZ), cold continental outbreaks, and the sporadic Caribbean cyclones [17–19]. Strong orographic effects are caused by a NW to SE mountain range (or cordillera) with a maximum elevation of 3820 m above sea level (m a.s.l.), which divides the country into the Caribbean and Pacific regions, each region having a distinct precipitation regime. In the Pacific region of Costa Rica, the dry season ranges from December to April and the wet season ranges from May to November. There is a secondary humidity gradient along the Pacific coast where wetness increases from north to south [20,21]. The observed cyclic deviations in the ocean-atmosphere domain can be described as "wet" and "dry" years throughout Costa Rica and are mainly linked to changes in the sea surface temperature (SST), especially the warm/cold El Niño Southern Oscillation (ENSO) episodes [17,22].

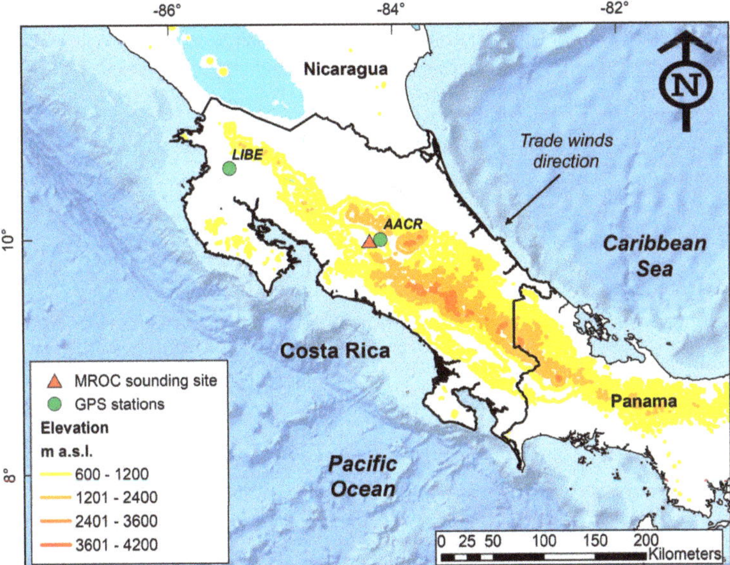

Figure 1. The location of the GPS stations (Liberia, LIBE and Central Valley, AACR, green circles) in Costa Rica and the atmospheric sounding site at the San José International Airport (International Civil Aviation Organization code: MROC, red triangle): The AACR and the MROC sounding site are situated in the central mountainous region of Costa Rica (Central Valley), whereas LIBE is located on the dry corridor of Central America (northern Pacific of Costa Rica).

2.2. GPS and Atmospheric Sounding Data

As stated above, there are 14 GNSS stations in operation in Costa Rica, which are associated with the SIRGAS network. We selected GPS data from two of these stations to estimate PWV: one located in the Central Valley (AACR) and one situated in the northern Pacific (Liberia or LIBE). As shown in Figure 1, AACR is located in the mountainous central region of the country known as the Central

Valley and LIBE is situated on the northern Pacific region. We selected these two stations based on three criteria: i) at least one station must be as near as possible to an atmospheric sounding site (AACR), ii) at least one additional station must be included in the analysis and situated in the Pacific slope of the country (the climatic region where the Central Valley is located, LIBE), and iii) for each station, a weather station must be available to register the meteorological conditions, with no significant height differences between the GPS station and the weather station.

The GPS data were processed by the National Processing GNSS Data Center. The receiver and antenna types at AACR were Topcon TPS NET-G3A and Topcon TPSCR.G3 TPSH, respectively. At LIBE, the receiver and antenna model were Leica GRX 1200 + GNSS and Leica AT504GG LEIS respectively. GPS data were processed using the software GIPSY, version 6.4 from JPL [23], using the Precise Positioning Point (PPP) method based on the precise ephemerides computed by JPL. The parameters for the satellite and receiver antenna phase center calibration were set according to the JPL products. The tropospheric model incorporated a priori hydrostatic delay (PHD, m), computed as follows:

$$\mathrm{PHD} = 1.013 * 2.27e^{(-0.00016*h)}, \tag{1}$$

where h is the station height (m) above the ellipsoid. The PHD value was estimated as 0.1 m. The tropospheric gradient was estimated based Bar-Sever et al. [24]. The global mapping function (GMF) troposphere mapping functions were implemented, and an elevation cutoff angle was set at 7.5°.

The observations of AACR and LIBE stations were available from January 1st to December 31st, 2017. To obtain PWV radiosonde estimations, we used the only atmospheric sounding site in operation in Costa Rica, namely the International Airport of San José, Costa Rica (International Civil Aviation Organization code: MROC). The radiosonde launching was carried out by the National Meteorological Institute of Costa Rica using mainly Sprenger E085 (St. Andreasberg, Germany) sounding systems. The radiosonde data were obtained from the University of Wyoming [25]. The distances and elevation differences between the GPS stations and the atmospheric sounding site are summarized in Table 1. In situ meteorological observations were measured with a Vantage Pro2 weather station (Davis Instruments, Hayward, CA, USA), with no significant height difference between the GPS stations and the weather monitoring sites.

Table 1. The location details for AACR and LIBE GPS receivers in the Central Valley and northern Pacific region of Costa Rica and for the MROC radiosonde site used in this study.

Station	AACR	LIBE	MROC
Latitude (decimal degrees)	9.9386	10.6305	9.9944
Longitude (decimal degrees)	−84.1179	−85.4380	−84.2079
Elevation (m a.s.l.)	1159	132	912
Δ distance (km) [1]	11.5	152	-
Δ elevation (m a.s.l.) [1]	+247	−780	-

[1] Δ the distance and Δ the elevation in relation to the MROC sounding site.

2.3. GPS Data Processing

In general, GPS data processing is based on the physics of the atmospheric propagation delay. GPS radio waves are delayed by the ionosphere and troposphere when they travel through the atmosphere from the satellite to GPS ground-receivers. The so-called "total or zenith atmospheric delay" (or ZTD, in millimeters) of the signal emitted by a GPS satellite consists of two parts, "hydrostatic delay" or ZHD and "wet delay" or ZWD:

$$\mathrm{ZTD} = \mathrm{ZHD} + \mathrm{ZWD} \tag{2}$$

Overall, the ZHD is due to the effect of dry air, contributing to at least 90% of the total tropospheric delay, whereas the ZWD represents less than 10% of the signal. Therefore, the ZTD depends on the air mass between the receiver and satellite and can be expressed as a function of ground atmospheric pressure [26–28]:

$$\mathrm{ZHD} = \frac{0.002277\ \mathrm{P_{surf}}}{1 - 0.00266\cos(2\ \theta) - 0.00028\mathrm{H_{site}}}, \tag{3}$$

where P_{surf} is the surface pressure (hPa), θ the geodetic latitude, and H_{site} represents the height (km) above the geoid [26]. Once the ZHD is calculated, ZWD is estimated by subtracting ZHD from ZTD.

Overall, the computation of ZWD using GPS delay is commonly related to the precipitable or integrated humidity along the altitudinal profile over the local atmosphere (known as GPS PWV). GPS PWV represents the total mass of water vapor in an atmospheric column with a unit area and is measured in kg/m^2, but it is usually reported as the height of an equivalent column of liquid water in millimeters [26,29]. We used the Bevis relationship to estimate GPS PWV using ZWD [8]:

$$\mathrm{PWV} = \mathrm{kZWD},\ \mathrm{k} = \frac{10^6}{\left(\frac{k_3}{T_m} + k'_2\right)R_v\rho}, \tag{4}$$

where k is a dimensionless water vapor conversion coefficient. In Equation (4), k_3 and k'_2 are empirical constants [30], Rv is the specific gas constant for water vapor, ρ is the liquid water density, and T_m is the mean temperature of the atmospheric column. To calculate T_m, we used the well-known Bevis equation [30]:

$$T_m = 70.2 + 0.72T_s, \tag{5}$$

where T_s is the surface air temperature. In order to test the validity of this relationship for the tropical atmosphere of Costa Rica, we used the radiosonde profiles available at MROC during the study period (N = 210) to estimate the T_m values for the local atmosphere. As T_m depends both on the temperature profile and the vertical distribution of water vapor, T_m was calculated using the following equation [31,32]:

$$T_m = \frac{\int_0^\infty \frac{P_w}{T}dz}{\int_0^\infty \frac{P_w}{T^2}dz}, \tag{6}$$

where P_w is the water vapor pressure and T is the air temperature.

Based on the atmospheric sounding data available at MROC during 2017, the composite atmospheric temperature profiles depict lapse rate changes for the tropopause and troposphere for the dry and wet seasons (Figure 2A). During the dry season, the mean lapse rate was −5.0 °C/km from the ground to the tropopause level (approx. 15–20 km), whereas during the wet season, the mean lapse rate was −4.8 °C/km. Tropospheric temperature variations (up to 15 km) were similar during the wet season (range: 198–296 K, mean: 256 ± 27 K) and the dry season (range: 199–297 K, mean: 258 ± 28 K). At MROC, the mean surface temperature (T_s) varied from 293–296 K during the dry season and from 289–296 K during the wet season. The corresponding T_m values were in the range 279–289 K (dry season) and in the range 277–302 K (wet season). When we fitted a straight line to the T_m data to obtain a T_m–T_s relationship for MROC (black line in Figure 2B), we found a poor Spearman's correlation between T_m and T_s for our data (r = 0.0257, $p > 0.05$). Therefore, we compared the relative bias of the Bevis equation (Equation 5, plotted as a red line in Figure 2B) to the T_m calculations from the radiosonde data. The mean relative bias using the Bevis equation was −0.009 ± 0.008 for the dry season estimations (N = 70) and 0.004 ± 0.009 for the wet season calculation (N = 140). We also calculated a RMSE of 3.50 K for the dry season, with a RMSE of 2.72 K for the wet season. The estimated relative biases are equivalent to the mean error values of −2.6 ± 2.4 K (dry season) and 1.1 ± 2.5 K (wet season). In terms of GPS PWV, the mean error associated to these T_m deviations were in the range of −0.2 and 0.4 mm. Therefore, as the relative biases for the Bevis equation are smaller than the estimated precision for the mean daily PWV calculations, we decided to apply the Bevis equation to estimate T_m in the calculations of PWV at our study sites.

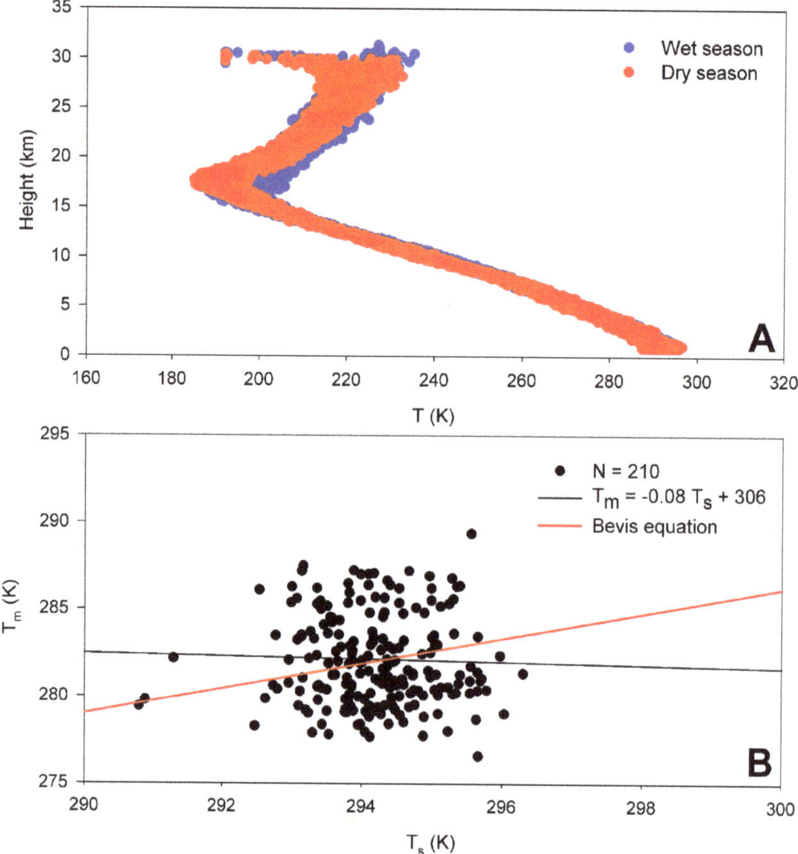

Figure 2. (**A**) A composite atmospheric temperature (K) profile constructed using radiosonde measurements at the MROC sounding site during the dry season (N = 70, red dots) and wet season (N = 140, blue dots). (**B**) The surface temperature (T_s, K) vs. mean temperature of the atmospheric column (T_m, K) used to calculate the T_m–T_s relationship for the Central Valley of Costa Rica (black line) and the Bevis equation [8] (red line).

In this work, we report mean daily PWV estimations based on hourly calculated ZTD and ZHD values at AACR and LIBE. We decided to carry out our analysis on a daily basis because there is limited atmospheric sounding data for Costa Rica, with only one radiosonde station in operation. Therefore, sounding data can be only considered representative of the average daily atmospheric conditions. The average precisions associated with the mean daily PWV calculations are 1.3 mm and 1.1 mm for AACR and LIBE, respectively.

2.4. MODIS Data

MODIS is a radiometer on board the Terra (launched in 1999) and Aqua (launched in 2002) satellite platforms. The MODIS instruments on the Terra and Aqua image the same area on Earth approximately three hours apart, observing the entire Earth's surface every 1 to 2 days. Terra's sun-synchronous, near-polar circular orbit passes the equator from north to south (descending node), whereas Aqua's sun-synchronous, near-polar circular orbit crosses the equator from south to north (ascending node). The water vapor remote sensing method is based on detecting the absorption by the water vapor of the reflected solar radiation after it has transferred down to the surface and back up through the atmosphere. The total vertical amount of water vapor can be estimated from a comparison between the reflected solar radiation in the absorption channel and the reflected solar radiation in nearby non-absorption channels. The solar radiation between 0.86 and 1.24 μm on the sun-surface-sensor path is subjected to atmospheric water vapor absorption but also to atmospheric aerosol scattering and surface reflection. Therefore, in order to estimate column water vapor from measurements of the solar radiation reflected by the surface, the absorption and scattering properties of the atmosphere and the surface near 1 μm must be considered [33]. The PWV products are derived from infrared (IR) and near-infrared (NIR) measurements. NIR bands are used for daytime measurements (solar radiation reflected by Earth + atmosphere), and IR bands are used during nighttime conditions (radiation emitted by Earth + atmosphere). If clouds are present, other channels in the range of the 0.8–2.5μm region can be used in order to estimate the absorption due to water vapor above and within clouds [13,14].

Among the available MODIS products, the Level-3 MODIS Atmosphere Daily Global Product contains roughly 600 statistical datasets that are derived from approximate 80 scientific parameters from four Level-2 MODIS Atmosphere Products: Aerosol, Water Vapor, Cloud, and Atmosphere Profile. There are two MODIS Daily Global data product files: MOD08_D3, containing data collected from the Terra platform, and MYD08_D3, containing data collected from the Aqua platform [11]. In this study, the level-3 MODIS Terra and Aqua products of the daily mean (MOD08_D3 and MYD08_D3, respectively) global grid with a spatial resolution of (1° × 1°) were used to conduct the GPS PWV comparison during 2017. We selected the square region of 30 × 30 km dimensions centered on the MROC sounding site to calculate the satellite PWV estimations [34]. MODIS data estimates were calculated as area-averaged values and were processed using the Earth Observing System Data and Information System (EOSDIS) Giovanni website [35]. A total of 299 and 267 PWV Aqua and Terra satellite estimations were available from the MODIS data product, respectively, for the study period. The typical uncertainty of the MODIS PWV estimations is approximately 5–10% [13].

2.5. HYSPLIT Air Mass back Trajectory Analysis

Air mass back trajectory analyses were conducted using the HYSPLIT Lagrangian model developed by the Air Resources Laboratory (ARL) of the National Oceanic and Atmospheric Administration (NOAA, USA) [36,37]. Representative air parcel trajectories were estimated 72 h backwards in time due to the nearness of the Caribbean Sea and the Pacific Ocean. Each trajectory was calculated using NOAA's meteorological data files (GDAS, global data assimilation system: 2006–present; 0.5° resolution) as input for the HYSPLIT model [38]. The ending altitude of air masses was set to the mean elevation of the Central Valley of Costa Rica (approx. 1,100 m a.s.l.). Trajectory analysis ending times at the Central Valley (AACR) were set to 12:00 UTC, which corresponds to a local time of 06:00 a.m. in Costa Rica. Given the estimated residence time of water in the atmosphere, ranging from around 4–10 days, weekly (N = 52, Figure 3), air mass back trajectories were calculated [38]. The ending dates for the trajectory analysis were set on Sunday of every week. These air masses were classified into two main groups, dry season (January–April) and wet season (May–December), to compare and identify the moisture transport pathways followed by the air masses that arrived at the Central Valley of Costa Rica.

2.6. Statistical Analysis

A Kruskal–Wallis non-parametric one-way analysis of variance on ranks was used to investigate if the GPS PWV stochastically dominates the other PWV estimations (i.e., atmospheric sounding and MODIS Aqua and Terra) during the dry and wet season [39]. A pairwise multiple comparison procedure was applied using Dunn's method for those groups having a significant difference in PWV in order to isolate the stochastic dominance of the group or groups that differ from the others [40].

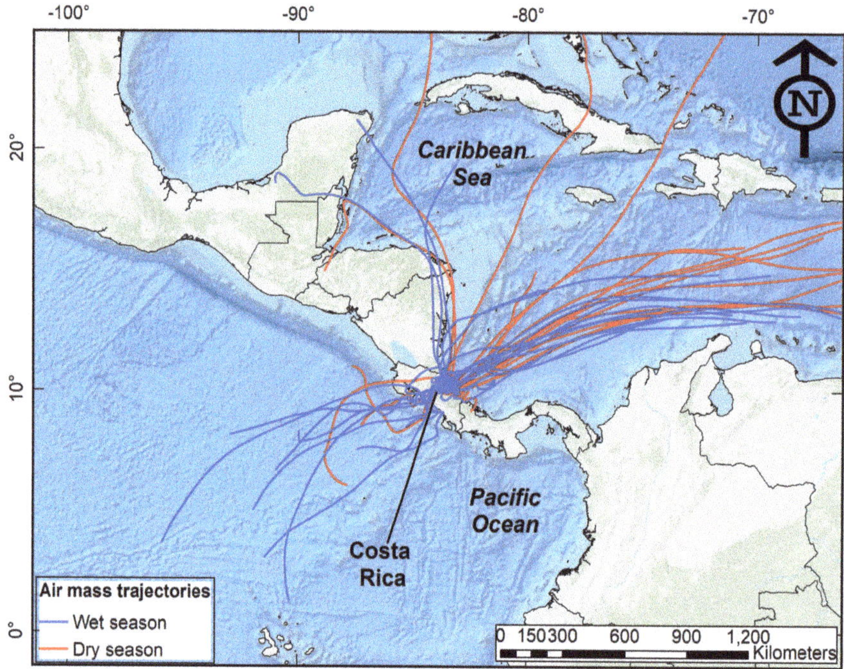

Figure 3. Representative 72-h air mass back trajectories for the dry (red) and wet (blue) seasons in 2017 calculated using the HYSPLIT Lagrangian model [26].

We also applied a multiple linear regression (MLR) model using surface meteorological data in order to identify the major variables controlling the PWV values in the Central Valley (AACR GPS station).

The cumulative annual precipitation for AACR during 2017 was 2586 mm with a mean daily precipitation during the dry season of 2 mm (range: 0 mm–54 mm) and 11 mm (range: 0 mm–85 mm) during the wet season. In LIBE, the corresponding cumulative annual precipitation was 2161 mm with a mean daily precipitation during the dry season of 1 mm (range: 0 mm–49 mm) and 9 mm (range: 0 mm–247 mm) during the wet season. At both sites, maximum daily precipitation values were recorded at the end of October (Figure 4A). Despite the differences in the cumulative annual precipitation, the relative humidity and air temperature variations were similar in the two regions. The mean relative humidity was 80.8 ± 5.6 % (range: 39.0%–92.6%) and 82.4 ± 8.0% (range: 34.6%–93.6%) in AACR and LIBE, respectively (Figure 4B). The mean air temperatures were 21.0 ± 0.9°C (range: 16.2 °C–23.2 °C) and 26.7 ± 1.3 °C (range: 22.5 °C–30.5 °C) in AACR and LIBE, respectively (Figure 4C). However, on a seasonal basis, the mean daily air temperatures and mean daily relative humidity were 6.3 °C and 2.1% greater in LIBE than in AACR during the dry season, respectively. During the wet season, the corresponding values were 5.2 °C and 1.2% greater in LIBE than in AACR, in that order.

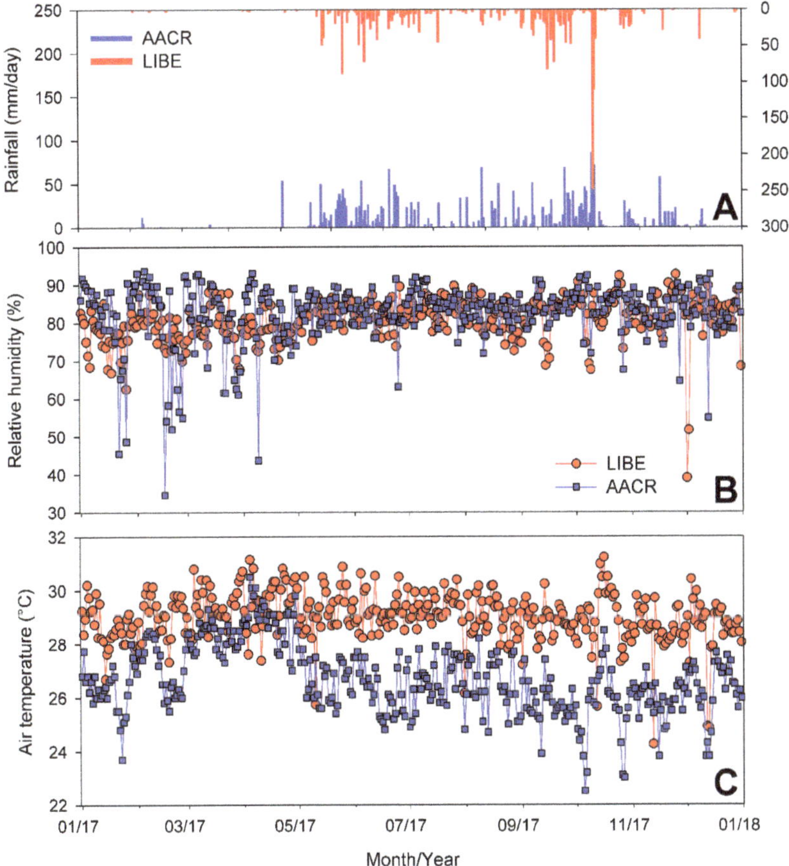

Figure 4. The time series of (**A**) the daily precipitation (mm/day) recorded in Heredia (blue bars) and in Liberia (red bars) during 2017: The left y-axis corresponds to Heredia, whereas the right y-axis shows the data of Liberia. (**B**,**C**) The average daily relative humidity (%) and air temperature (°C) for Heredia (blue circles) and Liberia (red circles).

3. Results

3.1. Seasonal Variations of GPS PWV in AACR and LIBE

During the study period, the HYSPLIT trajectory analyses identified that air masses reaching Costa Rica predominantly came from the southeastern Caribbean Sea with a less frequent contribution from the Pacific Ocean (Figure 3). Overall, the wind direction and speed in Costa Rica are mostly influenced by the seasonal migration of the ITCZ. Thus, during the dry season (December–April) when the ITCZ is located south of Costa Rica, air mass trajectories were associated with the influence of the NE trade winds. During the wet season (May–November), NE trade winds were weaker due to the passage of the ITCZ over Costa Rica, and cross-equatorial winds from the southern hemisphere transported moisture from the Pacific Ocean to the Central American Isthmus. This moisture transport pattern controlled the precipitation regimes observed at the Central Valley (AACR, Figure 4A) and the northern Pacific region of Costa Rica (LIBE, Figure 4A). During the study period, approx. 23% (N = 12) of the air masses arrived from the Pacific Ocean and the rest (approx. 77%, N = 40) came from the Caribbean Sea. Air masses arriving from the Pacific Ocean and the Caribbean Sea predominantly traveled over the eastern Pacific Ocean and the central and southern Caribbean Sea basins, respectively. No significant differences were found between the mean sea levels of the air masses that reached the Central Valley in the dry and wet seasons, with typical mean sea levels of 1500 m to 2000 m.

Seasonal GPS PWV variations were clearly defined at AACR and LIBE (Figure 5A). During the dry season, GPS PWV values varied from 14.8 mm to 40.9 mm in AACR (mean value: 27.6 ± 6.3 mm), whereas in LIBE, the variation was in the range 20.2 mm–55.5 mm (mean value: 36.9 ± 7.6 mm). We observed an increment in the GPS PWV values at both sites at the end of April and at the beginning of May that coincides with the onset of the wet season in Costa Rica, namely the passage of the ITCZ. During the wet season, at AACR, the GPS PWV estimates ranged from 24.3 mm to 46.2 (mean value: 39.7 ± 3.7 mm), and at LIBE, these GPS PWV values fluctuated from 31.5 mm to 62.6 mm (mean value: 54.1 ± 5.2 mm). At the end of November, when the transition wet-to-dry season began, we registered a decrease in the GPS PWV estimations related to the beginning of the dry season and the influence of the NE trade winds. Overall, the GPS PWV values were greater at LIBE than at AACR due to the elevation difference between the GPS stations (Δ1,027 m a.s.l.). For example, the mean differences between the estimations for AACR and LIBE were −9.5 ± 4.5 mm in the dry season and −14.4 ± 3.0 mm in the wet season. As shown in Figure 5A, these observed differences in the GPS PWV measurements at the GPS stations were more evident during the wet season when the ITCZ predominantly influenced the air circulation over Costa Rica. During the dry season, on the other hand, the differences between the GPS PWV values for AACR and LIBE were relatively more difficult to separate. However, although the GPS stations are situated approx. 160 km from each other (one in the Central Valley, AACR, and the other one in the northern Pacific region of Costa Rica, LIBE), we found a good Spearman's correlation ($r = 0.929$, $p > 0.001$) between the GPS PWV values estimated for AACR and the corresponding estimations calculated for LIBE (Figure 5B). The best performing linear regression model shown in Figure 5B overall explained 86.9% of the variance for the GPS PWV estimates calculated for LIBE using the GPS PWV values at AACR. Overall, this finding confirms that the PWV variations at both sites are controlled by the climatic conditions of the Pacific slope which is also reflected in the precipitation patterns and air temperature/relative humidity variations shown in Figure 4A–C. This is an important result that demonstrates the applicability of PWV to monitor changes in the hydrometeorological conditions at regions that share similar climatic conditions. Additionally, our HYSPLIT analysis is able to identify the seasonal PWV variations at AACR. For instance, air masses arriving from the Pacific Ocean between May and October are associated with high PWV estimations with values between 39 and 44 mm/day. These values are practically equal to or greater than the 75th percentile of our data set (41 mm/day). In turn, air masses coming from the Caribbean Sea were associated with greater variations in the PWV estimations registered between November and April but also to smaller PWV estimations (up to 14 mm/day, Figure 5A).

3.2. GPS PWV Comparison to Other Estimations Methods and MRL Analysis

GPS PWV observations at the Central Valley of Costa Rica (AACR) compared well to the atmospheric sounding measurements during the dry and wet season but only to the MODIS Terra estimations during the dry season. As shown in Figure 6A, GPS PWV observations followed the seasonal variations registered using the radiosonde data. The best performing satellite-based estimations were those retrieved from the MODIS Terra, which also followed the seasonal variations in the GPS and radiosonde PWV observations. Unlike MODIS Terra, MODIS Aqua PWV estimations showed a systematic positive bias with respect the GPS PWV values and the radiosonde data. To better identify the seasonal differences found after applying these PWV estimation methods, we split our data set into two groups: dry season and wet season estimations (Figure 6B,C). For the dry season, the GPS PWV median value (26.5 mm) was not significantly different from the radiosonde PWV median value and the MODIS Terra PWV median estimation (27.0 mm and 25.8 mm, respectively; $p > 0.05$). However, it was significantly different from the median value estimated using the MODIS Aqua PWV values (29.7 mm, $p > 0.001$). In turn, for the wet season, the GPS PWV median value (40.3 mm) was significantly different from the MODIS Terra and MODIS Aqua PWV median estimations (36.0 mm and 51.4 mm, respectively; $p < 0.001$) but not significantly different from the radiosonde PWV median value (41.4 mm, $p > 0.05$). The mean relative biases for MODIS Aqua PWV and MODIS Terra PWV were also calculated using the GPS PWV as a reference. During the dry season, these values corresponded to 0.16 ± 0.24 mm and 0.02 ± 0.30 mm, respectively, and were equivalent to RMSE values of 7.43 mm and 7.21 mm, in that order. During the wet season, the mean relative biases were 0.30 ± 0.24 mm and −0.06 ± 0.19 mm, respectively, corresponding to 15.2 mm and 8.05 mm, respectively.

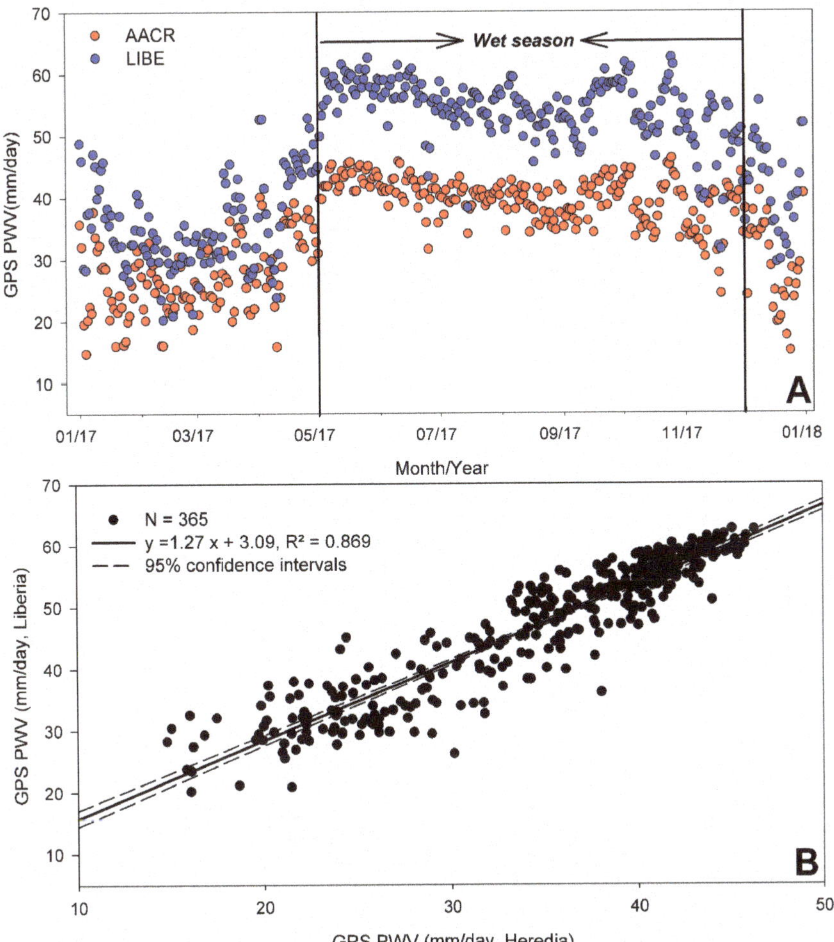

Figure 5. (**A**) The time series of GPS precipitable water vapor (PWV) (mm/day) estimated for AACR (blue circles) and LIBE (red circles): The wet season period 2017 (May–November) is delimited in the graph. (**B**) The graph shows the relationship between the GPS PWV estimated for ACCR and LIBE ($p < 0.001$, N = 365). The 95% confidence limits are also shown (dashed lines).

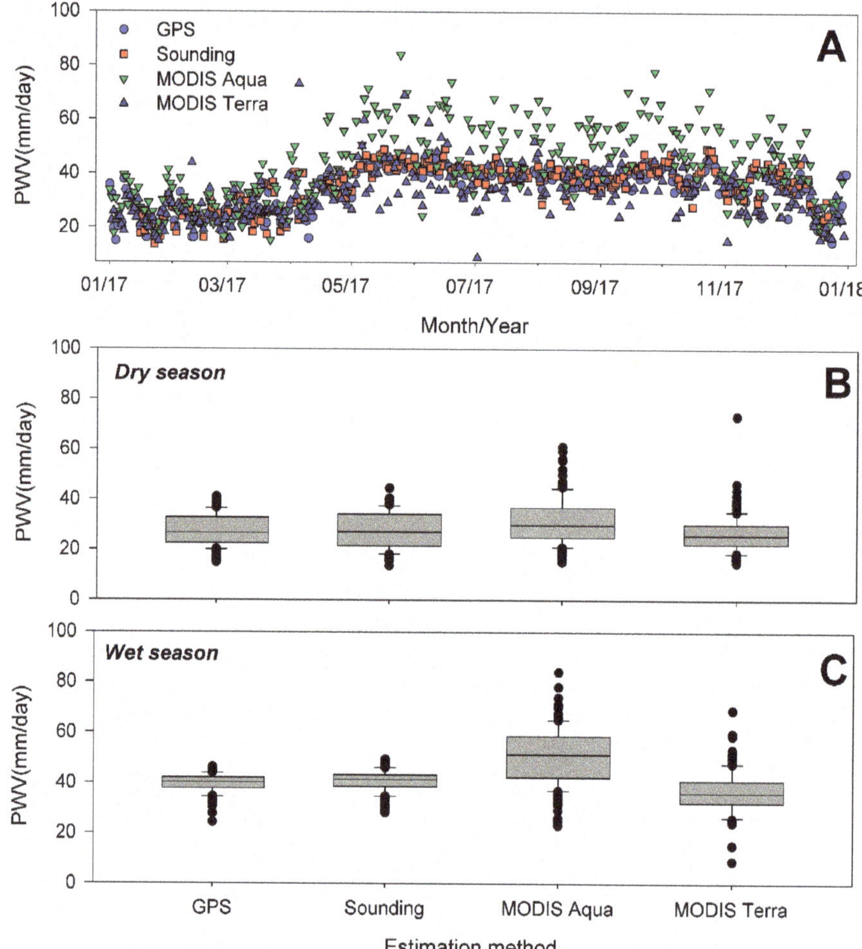

Figure 6. (**A**) The time series of PWV (mm/day) estimated for AACR using GPS (blue circles), atmospheric sounding (red squares), MODIS Aqua (green inverted triangles), and MODIS Terra (blue triangles). (**B**,**C**) Box plots of the PWV (mm/day) estimated using GPS, atmospheric sounding, MODIS Aqua, and MODIS Terra for the dry season and wet season in AACR, respectively: The grey box indicates the 25th and 75th percentiles with the median in middle. The error bars indicate the minimum and maximum values. The black circles indicate outliers (1.5 times the central box).

Using the available surface meteorological data at AACR, we conducted a multiple linear regression (MLR) analysis to identify the major drivers controlling the seasonal variability of GPS PWV measurements made at the Central Valley of Costa Rica (AACR, Figure 7A). The best performing MLR model was calculated as

$$\text{GPS PWV} = 4.257(\text{T}) + 0.355(\text{RH}) - 0.0486(\text{FLUX}) - 0.257(\text{P}) - 999.125,\ \text{R}^2 = 0.597 \quad (7)$$

where T is the mean daily air temperature (K), RH is the mean daily relative humidity (%), FLUX is the mean daily downward solar radiation flux (W/m^2), and P is the mean daily atmospheric pressure (hPa). The mean relative bias associated with the estimations of GPS PWV values using this model during the dry season was 0.10 ± 0.25 mm, whereas for the wet season the mean relative bias was

−0.04 ± 0.09 mm. The corresponding RMSE estimated for the dry and wet seasons were 6.09 mm and 4.02 mm, respectively. When this model was applied to estimate the sounding PWV measurements, the mean relative bias during the dry season and wet season were 0.12 ± 0.27 mm and −0.06 ± 0.10 mm, respectively. The RMSE values calculated for the dry and wet season were 6.83 mm and 4.72 mm, respectively. As shown in Figure 7B, the correlation between the GPS PWV data at MROC and the corresponding PWV values estimated from the MLR model was better for the values between 30 and 45 mm. As these values were mostly registered during the wet season, it seems that our MLR model performs better when the atmospheric conditions in the Central Valley are controlled by the seasonal migration of the ITCZ and worse during the less stable atmospheric conditions linked to the influence of NE trade winds.

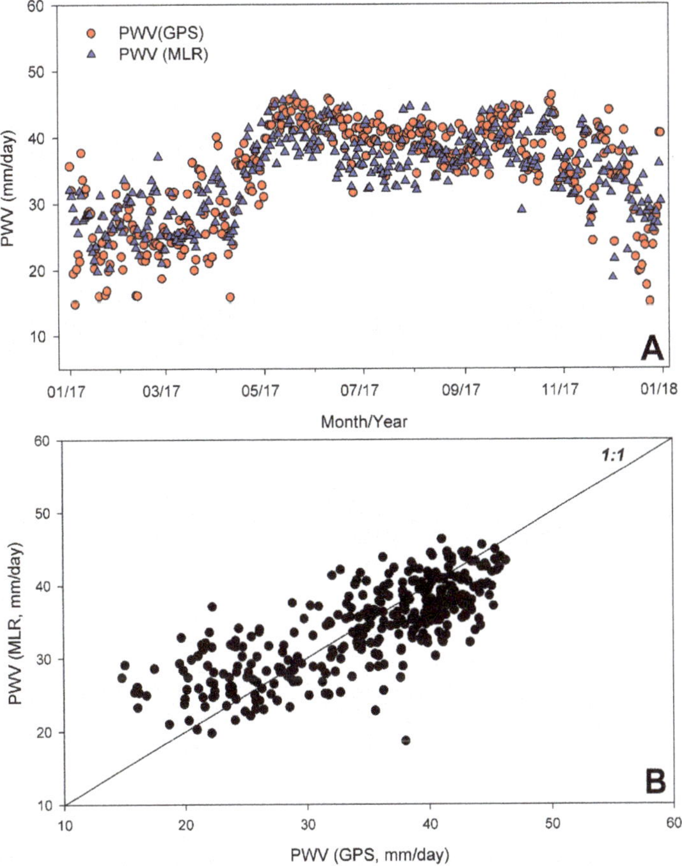

Figure 7. (**A**) The simulated PWV (mm/day) time series (blue triangles) in relation to the estimated GPS PWV (mm/day) at AACR (red circles). (**B**) The GPS PWV (mm/day) vs. simulated PWV at AACR shows the goodness-of-fit of the multiple linear regression (MLR) model.

4. Discussion

Because of the isthmian geographical environment of Costa Rica (with only a Pacific-to-Caribbean coast distance of approx. 400 km), the good correlation between the GPS PWV estimates at AACR and LIBE was an expected result of our analysis, as both sites are located on the Pacific slope and share similar climatological features (e.g., analogous precipitation patterns, Figure 4A). Moreover, we also confirm the good agreement between the GPS PWV calculations and the radiosonde-based measurements reported by others [6,9,26,27,41]. For example, Spearman's correlation coefficients for the GPS PWV and the radiosonde-based calculations were 0.913 ($p < 0.001$) and 0.902 ($p < 0.001$) during the dry and wet season, respectively. With respect to the MODIS satellite estimations of PWV, our analysis yielded significant biases depending on the season of the year, which are related to the annual cycle of water vapor, the NE trade winds influence, and the ITCZ activity over the Central American Isthmus. For instance, only the MODIS Terra PWV estimations recorded during the dry season were not significant biased with respect the GPS PWV calculations. However, we also found good correlations between the MODIS Aqua and the GPS-based calculations, with Spearman's correlation coefficients of 0.735 ($p < 0.001$) and 0.621 ($p < 0.001$) for the dry and wet season, respectively. The dry and wet season MODIS Terra Spearman's correlation coefficients were 0.591 ($p < 0.001$) and 0.368 ($p < 0.001$), respectively. These correlation values were similar to those reported over different regions of the Iberian Peninsula, including island environments like Mallorca and several coastal sites [9]. Such relationships also allowed a further adjustment of the data to fit the observations by adopting a spatial bias (error) correction method like the one applied to precipitation data [21,42,43]. As mentioned above, due to the location of Costa Rica on the narrow land-bridge of Central America, the MODIS near-infrared water vapor retrieval algorithm could be greatly affected and the derived column water vapor values over coastal or water areas may vary significantly due to the lower signal-to-noise ratios of the measured spectra [13]. This effect on the MODIS retrieval algorithm was particularly evident in the MODIS Aqua PWV estimation which was somewhat high during both the dry and wet season. MODIS Terra also showed deviations but were related to an underestimation during the wet season which can be related to the so-called shielding effect (i.e., clouds are probably occulting water vapor underneath them) [13,44]. The differences between the MODIS Aqua and MODIS Terra estimations could be attributed to their different passing times over the Central American Isthmus (MODIS Aqua crosses the equator in the afternoon, whereas MODIS Terra does it in the morning) and to the use of different radiations to estimate the water vapor during the day and night. The MODIS Aqua estimations could be higher than the corresponding MODIS Terra values because the algorithm uses IR radiation during nighttime, which could be affected by the presence of clouds with water vapor, leading to overestimations. Overall, our HYSPLIT air mass trajectory analysis is consistent with the prevailing regional moisture transport mechanism during the dry season, the Caribbean Low Level Jet (CLLJ). During the wet season, in turn, there is an intensification in the genesis and development of deep convection systems on the Pacific coast of Costa Rica which is generally is associated with the presence of the "Chorro del Occidente Colombiano" or CHOCO jet [45]. These circulation patterns produced the two rainfall maxima observed on the Pacific slope, one in June and one in September, which were interrupted by a relative minimum between July–August, known as the Midsummer Drought, due to the intensification of trade winds over the Caribbean Sea [46]. The radiosonde data were also useful to validate the atmospheric conditions controlling the GPS PWV estimations. First, the composite temperature profiles calculated using the radiosonde data are in agreement with the previously reported structure of the upper troposphere and lower stratosphere over Costa Rica [47]. As shown in Figure 2A, the temperatures in both the dry and wet season are roughly the same at 25 km, but below this level (e.g., 15–20 km), the boreal winter (December to April) temperature profile is colder than in boreal summer (from May to October). This finding was previously attributed to the influence of wave-induced vertical motions across strong vertical gradients, the source variability in the air masses arriving at Costa Rica (e.g., tropical western Pacific or midlatitudes) resulting from horizontal transport and changes induced along parcel paths due to physical and/or chemical processes [47–49]. Secondly, despite these differences in the thermal structure of the tropical atmosphere of Costa Rica, the T_m calculations using the Bevis equation showed small differences with respect to the corresponding calculations based on radiosonde data. This finding also agrees with the calculations made in Algeria and Argentina where Namaoui et al. and Fernández et al. estimated the uncertainty of the T_m values and found that variations up to 15K produced small differences in the final estimation of GPS PWV, which did not exceed 2 mm [27,40]. Thirdly, the poor correlation observed between T_s and T_m at MROC deserves further discussion. It is generally considered that the most accurate method to obtain T_m is by using both temperature and humidity profiles from radiosonde data [42,50]. Therefore, we have confidence that our T_m estimations are good approximations of the temperature profiles over the Central Valley of Costa Rica. A possible explanation for this finding is the mountainous and isthmian characteristics of the Costa Rica territory. The atmospheric sounding site is located on the southwestern area of the Central Valley. From this site, the distance to the Pacific coast is only 55–60 km. Radiosondes typically head in that direction after they are launched. Therefore, it seems that the atmospheric profiles estimated from MROC are representative not only of atmospheric conditions over the Central Valley but also of the Pacific coast of Costa Rica. There is also a limitation regarding the time of day when the sounding is performed. At MROC, atmospheric sounding is only done once a day, typically at 12Z or 7:00 a.m. Central American time. Therefore, T_m estimations with respect to T_s represent only the atmospheric conditions prevailing during the morning when constant surface temperatures are observed (approx. 294K ± 1K; Figure 2B). In consonance with these results, it was decided to rely on the Bevis equation to estimate the hourly T_m values for the Pacific slope of Costa Rica as this model has been extensively applied to estimate weighted atmospheric temperatures in several regions.

The MLR model estimated for GPS PWV data at AACR clearly matched the seasonal changes correctly, simulating smaller GPS PWV values during the dry season (from December to April) and much greater values during the wet season (from May to November). The best-performing and most parsimonious model included, as expected, near-surface (T and RH, Equation 7) and vertical atmospheric predictor variables (FLUX and P, Equation 6). The GPS PWV values were positively correlated with air temperature (T) and relative humidity (RH), with Spearman's correlation coefficients of 0.210 and 0.426 ($p < 0.001$), respectively, and were negatively correlated with solar radiation (FLUX) and air pressure (P), with Spearman's correlation coefficients of −0.360 and −0.175 ($p < 0.001$), respectively. These correlation results can be considered physically meaningful and can explain the overall model performance, although it is worth mentioning that, like the MODIS satellite estimations, it suffers from seasonal biases, specially during the dry season when the small PWV measured by the ground GPS receivers were not reproduced. This worse performance of the model during the dry season compared to the wet season was also evident after biases and RMSE values were additionally estimated using the sounding PWV measurements.

5. Conclusions

The combined analysis of PWV using GPS-based estimations, MODIS satellite products, and atmospheric sounding in the Pacific region of Costa Rica provides the first comparison between different water vapor calculation techniques for the Central American region. The evaluation of GPS-based estimates of PWV confirms the good performance of these estimations in comparison to the traditional and standard technique based on radiosondes, with no significant differences during the dry and wet seasons. These first results demonstrate the feasibility of incorporating GPS-based meteorological applications in order to improve the study of moisture inputs on the seasonal water vapor distribution in Central America. However, the performed evaluation identified significant biases between the GPS PWV estimates and the MODIS Aqua PWV estimations under both dry and wet season conditions and only the MODIS Terra PWV estimations recorded during the dry season were not significantly biased relating to the GPS PWV calculations. These results open the opportunity to evaluate other satellite products that provide higher spatial and temporal resolutions in order to give better insights into the causes of disagreements. Our analysis was also able to identify the influence of the main circulation patterns in Costa Rica, namely the trade wind regime and the ITCZ passage on PWV variability, which resulted in the relatively greater variability of the smaller PWV values during the dry season in comparison to the relatively smaller variability of the greater PWV values observed during the wet season. The influence of these moisture transport patterns was identified using the HYSPLIT analysis done for the Central Valley of Costa Rica. The multiple linear regression model successfully applied to this region can simulate the seasonal PWV variations using major meteorological variables, namely the mean daily air temperature, the mean daily relative humidity, the mean daily downward solar radiation flux, and the mean daily atmospheric pressure. We consider that a further analysis based on hourly GPS data could better analyze these relations between water vapor and HYSPLIT calculations and could refine the mathematical modeling presented in this work.

Author Contributions: Methodology, G.E.-H., J.F.V-C., and J.M.-Z.; software, J.F.V-C., S.R.-R., and J.M.-Z.; validation, P.C.-A. and S.R.-R.; writing—original draft, P.C.-A. and G.E.-H.; writing—review and editing, R.S.-M. and J.B.

Funding: This research received no external funding.

Acknowledgments: Germain Esquivel-Hernández, José Francisco Valverde-Calderón, and Ricardo Sánchez-Murillo thank the Research Office of the National University of Costa Rica through Grant SIA-0457-16.

Conflicts of Interest: The authors declare no conflict of interest.

References

1. Trenberth, K.E.; Fasullo, J.; Smith, L. Trends and variability in column-integrated atmospheric water vapor. *Clim. Dyn.* **2005**, *24*, 741–758. [CrossRef]
2. Kumar, S.; Allan, R.P.; Zwiers, F.; Lawrence, D.M.; Dirmeyer, P.A. Revisiting trends in wetness and dryness in the presence of internal climate variability and water limitations over land. *Geophys. Res. Lett.* **2015**, *42*, 10867–10875. [CrossRef]
3. Bosilovich, M.G.; Robertson, F.R.; Takacs, L.; Molod, A.; Mocko, D. Atmospheric Water Balance and Variability in the MERRA-2 Reanalysis. *J. Clim.* **2017**, *30*, 1177–1196. [CrossRef]
4. Sherwood, S.C.; Roca, R.; Weckwerth, T.M.; Andronova, N.G. Tropospheric Water Vapor, Convection, and Climate. *Rev. Geophys.* **2010**, *48*, RG2001. [CrossRef]
5. Giorgi, F. Climate change hot-spots. *Geophys. Res. Lett.* **2006**, *33*, L08707. [CrossRef]
6. Lu, N.; Qin, J.; Yang, K.; Gao, Y.; Xu, X.; Koike, T. On the use of GPS measurements for moderate resolution imaging spectrometer precipitable water vapor evaluation over southern Tibet. *J. Geophys. Res.* **2011**, *116*, 1–7. [CrossRef]
7. Jakobson, E.; Ohvril, H.; Elgered, G. Diurnal variability of precipitable water in the Baltic Region, impact on transmittance of the direct solar radiation. *Boreal Environ. Res.* **2008**, *14*, 45–55.
8. Bevis, M.; Businger, S.; Herring, T.; Rocken, C.; Anthes, R.; Ware, R. GPS Meteorology: Remote sensing of atmospheric water vapor using the global positioning system. *J. Geophys. Res.* **1992**, *97*, 15787–15801. [CrossRef]
9. Vaquero-Martínez, J.; Antón, M.; Ortiz de Galisteo, J.; Cachorro, V.; Costa, M.; Román, R.; Bennouna, Y. Validation of MODIS integrated water vapor product against reference GPS data at the Iberian Peninsula. *Int. J. Appl. Earth Obs.* **2017**, *63*, 214–221. [CrossRef]
10. Koulali, A.; Ouazar, D.; Bock, O.; Fadil, A. Study of seasonal-scale atmospheric water cycle with ground-based GPS receivers, radiosondes and NWP models over Morocco. *Atmos. Res.* **2012**, *104–105*, 273–291. [CrossRef]
11. MODIS Atmosphere. Available online: https://modis-atmosphere.gsfc.nasa.gov/products/daily (accessed on 15 January 2019).
12. Gui, K.; Che, H.; Chen, Q.; Zeng, Z.; Liu, H.; Wang, Y.; Zheng, Y.; Sun, T.; Liao, T.; Wang, H.; et al. Evaluation of radiosonde, MODIS-NIR-Clear, and AERONET precipitable water vapor using IGS ground-based GPS measurements over China. *Atmos. Res.* **2017**, *197*, 461–473. [CrossRef]
13. Gao, B.C.; Kaufman, Y.J. Water vapor retrievals using moderate resolution imaging spectroradiometer (MODIS) near-infrared channels. *J. Geophys. Res. Atmos.* **2003**, *108*, 1007–1021. [CrossRef]
14. Vaquero-Martínez, J.; Antón, M.; Ortiz de Galisteo, J.P.; Cachorro, V.E.; Álvarez-Zapatero, P.; Román, R.; Loyola, D.; Costa, M.J.; Wang, H.; González Abad, G.; et al. Inter-comparison of integrated water vapor from satellite instruments using reference GPS data at the Iberian Peninsula. *Remote Sens. Environ.* **2018**, *204*, 729–740. [CrossRef]
15. Diedrich, H.; Wittchen, F.; Preusker, R.; Fischer, J. Representativeness of total column water vapour retrievals from instruments on polar orbiting satellites. *Atmos. Chem. Phys.* **2016**, *16*, 8331–8339. [CrossRef]

16. Nilsson, T.; Böhm, J.; Wijaya, D.D.; Tresch, A.; Nafisi, V.; Schuh, H. Path Delays in the Neutral Atmosphere. In *Atmospheric Effects in Space*; Böhm, J., Schuh, H., Eds.; Springer: Berlin, Germany, 2013; pp. 73–129.
17. Waylen, M.E. Interannual variability of monthly precipitation in Costa Rica. *J. Clim.* **1996**, *9*, 2607–2613. [CrossRef]
18. Hidalgo, H.G.; Amador, J.A.; Alfaro, E.J.; Quesada, B. Hydrological climate change projections for Central America. *J. Hydrol.* **2013**, *495*, 94–112. [CrossRef]
19. Saénz, F.; Durán-Quesada, A.M.A. Climatology of low level wind regimes over Central America using a weather type classification approach. *Front. Earth Sci.* **2015**, *3*, 1–18. [CrossRef]
20. Powell, G.V.N.; Barborak, J.; Rodriguez, S.M. Assessing representativeness of protected natural areas in Costa Rica for conserving biodiversity: A preliminary gap analysis. *Biol. Conser.* **2000**, *93*, 35–41. [CrossRef]
21. Esquivel-Hernández, G.; Sánchez-Murillo, R.; Birkel, C.; Good, S.P.; Boll, J. Hydroclimatic and ecohydrological resistance/resilience conditions across tropical biomes of Costa Rica. *Ecohydrology* 2017. [CrossRef]
22. Alfaro, E.J. Some Characteristics of the Annual Precipitation Cycle in Central America and their Relationships with its Surrounding Tropical Oceans. *Top. Meteor. Ocean.* **2002**, *9*, 103.
23. Jet Propulsion Laboratory; California Institute of Technology. The Automatic Precise Positioning Service of the Global Differential GPS (GDGPS) System. Available online: http://apps.gdgps.net (accessed on 25 February 2018).
24. Bar-Sever, Y.E.; Kroger, P.M.; Borjesson, J.A. Estimating horizontal gradients of tropospheric path delay with a single GPS receiver. *J. Geophys. Res.* **1998**, *103*, 5019–5035. [CrossRef]
25. Atmospheric soundings. Available online: http://weather.uwyo.edu/upperair/sounding.html (accessed on 29 February 2018).
26. Benevides, P.; Catalao, J.; Miranda, P.M.A. On the inclusion of GPS precipitable water vapor in the nowcasting of rainfall. *Nat. Hazards Earth Syst. Sci.* **2015**, *15*, 2605–2616. [CrossRef]
27. Namaoui, H.; Kahlouche, S.; Belbachir, A.; Malderen, R.; Brenot, H.; Pottiaux, E. Water vapor and its comparison with radiosonde and ERA—Interim Data in Algeria. *Adv. Atmos. Sci.* **2017**, *34*, 623–634. [CrossRef]
28. Saastamoinen, J. Atmospheric Correction for the Troposphere and Stratosphere in Radio Ranging Satellites. In *The Use of Artificial Satellites for Geodesy*; Henriksen, S.W., Mancini, A., Chovitz, B.H., Eds.; The American Geophysical Union: Washington, DC, USA, 1972; Volume 15, pp. 247–251.
29. Wang, H.; Wei, M.; Li, G.; Zhou, S.; Zeng, Q. Analysis of precipitable water vapor from GPS measurements in Chengdu region: Distribution and evolution characteristics in autumn. *Adv. Space Res.* **2013**, *52*, 656–667. [CrossRef]
30. Bevis, M.; Businger, S.; Chiswell, S.; Herring, T.A.; Anthes, R.A.; Rockend, C.; Ware, R.H. GPS meteorology: Mapping zenith wet delays onto precipitable water. *J. Appl. Meteor.* **1994**, *33*, 379–386. [CrossRef]
31. Davis, J.L.; Herring, T.A.; Shapiro, I.I.; Rogers, A.E.E.; Elgered, G. Geodesy by radio interferometry: Effects of atmospheric modeling errors on estimates of baseline length. *Radio Sci.* **1985**, *20*, 1593–1607. [CrossRef]
32. Ross, R.J.; Rosenfeld, S. Estimating mean weighted temperature of the atmosphere for Global Positioning System applications. *J. Geophys. Res.* **1997**, *102*, 21719–21730. [CrossRef]
33. Atmosphere Discipline Team Imager Products. Water Vapor Algorithm Overview. Available online: https://modis-atmos.gsfc.nasa.gov/products/water-vapor/algorithm-overview (accessed on 3 April 2019).
34. Mooney, P.A.; Mulligan, F.J.; Fealy, R. Comparison of ERA—40, ERA—Interim and NCEP/NCAR reanalysis data with observed surface air temperatures over Ireland. *J. Clim.* **2011**, *31*, 545–557. [CrossRef]
35. Giovanni. The Bridge between Data and Science (v 4.28). Available online: https://giovanni.gsfc.nasa.gov/giovanni (accessed on 22 February 2018).
36. Stein, A.F.; Draxler, R.R.; Rolph, G.D.; Stunder, B.J.B.; Cohen, M.D. NOAA's HYSPLIT Atmospheric Transport and Dispersion Modeling System. *Bull. Am. Meteor. Soc.* **2015**, *96*, 2059–2077. [CrossRef]
37. Su, L.; Yuan, Z.; Fung, J.C.H.; Lau, A.K.H. A comparison of HYSPLIT backward trajectories generated from two GDAS datasets. *Sci. Total Environ.* **2015**, *506–507*, 527–537. [CrossRef]
38. van der Ent, R.J.; Tuinenburg, O.A. The residence time of water in the atmosphere revisited. *Hydrol. Earth Syst. Sci.* **2017**, *21*, 779–790. [CrossRef]
39. Kruskal, W.H.; Wallis, W.A. Use of ranks in one-criterion variance analysis. *J. Am. Stat. Assoc.* **1952**, *47*, 583–621. [CrossRef]
40. Dunn, O.J. Multiple comparisons among means. *J. Am. Stat. Assoc.* **1961**, *56*, 52–64. [CrossRef]

41. Fernández, L.I.; Salio, P.; Natali, M.P.; Meza, A.M. Estimation of precipitable water vapour from GPS measurements in Argentina: Validation and qualitative analysis of results. *Adv. Space Res.* **2010**, *46*, 879–894. [CrossRef]
42. Shi, F.; Xin, J.; Yang, L.; Cong, Z.; Liu, R.; Ma, Y.; Wang, Y.; Lu, X.; Zhao, L. The first validation of the precipitable water vapor of multisensor satellites over the typical regions in China. *Remote Sens Environ.* **2018**, *206*, 107–122. [CrossRef]
43. Vernimmen RR, E.; Hooijer, A.; Mamenun, A.E.; van Dijk AI, J.M. Evaluation and bias correction of satellite rainfall data for drought monitoring in Indonesia. *Hydrol. Earth Syst. Sci.* **2012**, *16*, 133–146. [CrossRef]
44. Román, R.; Antón, M.; Cachorro, V.E.; Loyola, D.; Ortiz de Galisteo, J.P.; de Frutos, A.; Romero Campos, P.M. Comparison of total water vapor column from GOME-2 on MetOp-A against ground-based GPS measurements at the Iberian Peninsula. *Sci. Total Environ.* **2015**, *533*, 317–328. [CrossRef]
45. Durán-Quesada, A.M.; Gimeno, L.; Amador, J.A.; Nieto, R. Moisture sources for Central America: Identification of moisture sources using a Lagrangian analysis technique. *J. Geophys. Res.* **2010**, *115*, D05103.
46. Magaña, V.; Amador, J.A.; Medina, S. The midsummer drought over Mexico and Central America. *J. Clim.* **2010**, *12*, 1577–1588. [CrossRef]
47. Schoeberl, M.R.; Selkirk, H.B.; Vömel, H.; Douglass, A.R. Sources of seasonal variability in tropical upper troposphere and lower stratosphere water vapor and ozone: Inferences from the Ticosonde data set at Costa Rica. *J. Geophys. Res. Atmos.* **2015**, *120*, 9684–9701. [CrossRef]
48. Selkirk, H.B.; Vömel, H.; Valverde, J.M.; Pfister, L.; Diaz, J.A.; Fernández, W.; Amador, J.; Stolz, W.; Peng, G.S. Detailed structure of the tropical upper troposphere and lower stratosphere as revealed by balloon sonde observations of water vapor, ozone, temperature, and winds during the NASA TCSP and TC4 campaigns. *J. Geophys. Res.* **2010**, *115*, D00J19. [CrossRef]
49. Fujiwara, M.; Vömel, H.; Hasebe, F.; Shiotani, M.; Ogino, S.-Y.; Iwasaki, S.; Nishi, N.; Shibata, T.; Shimizu, K.; Nishimoto, E.; et al. Seasonal to decadal variations of water vapor in the tropical lower stratosphere observed with balloon-borne cryogenic frost point hygrometers. *J. Geophys. Res.* **2010**, *115*, D18304. [CrossRef]
50. Wang, X.; Zhang, K.; Wu, S.; Fan, S.; Cheng, Y. Water vapor-weighted mean temperature and its impact on the determination of precipitable water vapor and its linear trend. *J. Geophys. Res. Atmos.* **2016**, *121*, 833–852. [CrossRef]

Article

A Lagrangian Ocean Model for Climate Studies

Patrick Haertel

Department of Geology and Geophysics, Yale University, New Haven, CT 06511, USA; patrick.haertel@yale.edu

Received: 1 February 2019; Accepted: 9 March 2019; Published: 15 March 2019

Abstract: Most weather and climate models simulate circulations by numerically approximating a complex system of partial differential equations that describe fluid flow. These models also typically use one of a few standard methods to parameterize the effects of smaller-scale circulations such as convective plumes. This paper discusses the continued development of a radically different modeling approach. Rather than solving partial differential equations, the author's Lagrangian models predict the motions of individual fluid parcels using ordinary differential equations. They also use a unique convective parameterization, in which the vertical positions of fluid parcels are rearranged to remove convective instability. Previously, a global atmospheric model and basin-scale ocean models were developed with this approach. In the present study, components of these models are combined to create a new global Lagrangian ocean model (GLOM), which will soon be coupled to a Lagrangian atmospheric model. The first simulations conducted with the GLOM examine the contribution of interior tracer mixing to ocean circulation, stratification, and water mass distributions, and they highlight several special model capabilities: (1) simulating ocean circulations without numerical diffusion of tracers; (2) modeling deep convective transports at low resolution; and (3) identifying the formation location of ocean water masses and water pathways.

Keywords: Lagrangian numerical method; ocean modeling; ocean mixing

---

1. Introduction

Many aspects of geophysical fluid dynamics are most easily modeled in a frame of reference that moves along with the fluid. For example, we show below that in such a framework three relatively simple ordinary differential equations predict large-scale fluid motions and local changes to fluid tracers (Section 2.3). While *partially* Lagrangian models, in which fluid particles are tracked in flow fields simulated by Eulerian methods, are becoming increasingly popular for both the atmosphere and the oceans [1–3], few *fully* Lagrangian general circulation models have been used for either the oceans or the atmosphere. This paper discusses the development of a fully Lagrangian model for simulating global ocean circulations.

One particular application the author has in mind for this model, is to couple it to his fully Lagrangian atmospheric model (LAM) [4]. The resulting Lagrangian coupled model is expected to be useful for weather and climate prediction as well as climate dynamics experiments. The LAM has already had success in simulating the Madden Julian Oscillation (MJO) [4–6], a planetary scale tropical weather disturbance [7] that is poorly represented in many atmospheric models [8–10], and which has global impacts on weather and climate [11–13]. The MJO also represents a component of the global atmospheric circulation that is predictable at much longer time scales than mid-latitude baroclinic waves [14]. The LAM has a unique spherical geometry [4], which as we note below is shared by the new global Lagrangian ocean model (GLOM), which makes the GLOM well suited for coupling to the LAM. However, there are many other potential advantages and applications for a fully Lagrangian global ocean model.

One such advantage is that the GLOM can circulate water without producing numerical tracer diffusion. A tracer value for a given water parcel does not change unless a tracer source is included, or a parameterization of tracer mixing is applied [15]. This contrasts with the behavior of z-coordinate ocean models, which generate an uncertain amount of spurious numerical mixing associated with the advection of tracers [16]. While isopycnal models [17] remove spurious numerical mixing of tracers *across* isopycnal surfaces—essentially by using a Lagrangian vertical coordinate—they also generate numerical mixing of tracers in modeling flow *along* isopycnals.

Another advantage of Lagrangian modeling relevant to this study relates to the parameterization of convection. In nature, when a fluid is unstable, convective plumes form that transport dense fluid downward and less dense fluid upward. The horizontal scale of the convective circulations is typically much smaller than that of the grid spacing for a large-scale model. What geophysical convective plumes ultimately accomplish is transporting fluid to a different layer. For example, in the atmosphere, tropical deep convection moves air from the boundary layer to the upper troposphere [18], and mid-level air to the boundary layer [19]. In the ocean, deep convective plumes move dense water from near the surface to the ocean bottom. As noted in Section 2.2, the Lagrangian convective parameterization produces the same vertical transport—by changing the vertical positioning of parcels—without attempting to model the details of the small-scale circulation in the convective plumes. The author believes that this unique convective parameterization has helped the LAM to simulate MJOs with realistic vertical structures and life cycles [4–6].

A third advantage of the Lagrangian approach to weather and climate modeling is that it provides fluid trajectory information for every mass element in the oceans and the atmosphere. Each component of fluid mass has an identification number that does not change during the course of a simulation [4,15]. The modeler can look up the position of a given mass element at all previous times for which model data is saved with no additional computations. It is hard to overemphasize the utility of this information. For example, it is well known that in the ocean, water mass characteristics are intimately connected to the locations in which they form, which is why water masses are given names like "North Atlantic Deep Water", "Antarctic Bottom Water", and "Antarctic Intermediate Water". The Lagrangian ocean modeler can easily determine where each water parcel last had contact with the surface (e.g., see Section 3.5). Similarly, in the atmosphere, the air temperature and moisture of a given air parcel are strongly dependent on previous locations of the parcel, with terms like "Artic Air" or "Gulf Moisture" (indicated moisture coming from the Gulf of Mexico) frequently used by meteorologists, and transports of moisture largely determining where moist convective systems form and move [4].

Finally, we note that perhaps the most important benefit of developing fully Lagrangian weather and climate modeling components, is that they will increase the genetic diversity of models. By perturbating the planet's radiative forcing, humans are essentially using the earth as a laboratory for a giant climate experiment with unknown consequences. We need a diversity of tools to map out the possible impacts of this experiment. Much in the same way that different kinds of engines (e.g., standard combustion, diesel, electric motors) have different niches for which they are best suited, so it is for different kinds of models of the atmosphere and oceans. The models and parameterizations that are the most fundamentally different from the others make the largest contribution to diversity. The Lagrangian tools presented here fit into that category. Much in the same way that the unique animals living in Australia are a great treasure to the biologist, Lagrangian modeling components are unique, and they have the potential to provide a fresh and fundamentally different perspective on ocean and atmosphere modeling and dynamics.

The work presented in this paper builds on a number of previous studies involving Lagrangian models of lakes and oceans. Haertel and Randall [20] developed a Lagrangian numerical method for the oceans in which a body of water was represented as a collection of conforming water parcels referred to as "slippery sacks". This method was closely related to two Lagrangian numerical methods used at that time: partical in cell (PIC; [21]) and smoothed particle hydrodynamics (SPH; [22]). However, in geophysical applications, PIC had only been used for one or two layers of fluid (e.g., [23]), and the slippery sacks method differed from SPH in that two conforming water parcels could not occupy the same physical space whereas two SPH particles could. Consequently, as slippery sacks moved around, they maintained a fairly uniform distribution over time, whereas regions highly concentrated with particles or nearly void of particles could develop under SPH. Haertel et al. [24] adapted horizontal and vertical mixing schemes to the slippery sacks method and created a three-dimensional model of a large lake. Haertel et al. [25] developed an idealized Lagrangian model of the North Atlantic Ocean. Haertel and Fedorov [15] added an Antarctic circumpolar to the idealized Atlantic Ocean Model. Applications of these models have included large lake upwelling [24], simulation of Stommel and Munk solutions [25], modeling of the circulation and stratification in the North Atlantic Ocean [15,25,26], and simulations of idealized equatorial oceans and tropical instability waves [27].

This study advances our previous Lagrangian ocean modeling efforts in several ways. In particular, for the first time global spherical geometry and realistic bottom topography are included in an ocean model. This means that for the first time a fully Lagrangian ocean general circulation model is presented that can be used in global weather and climate modeling and climate dynamics experiments. This model (the GLOM) was developed by combining components of the Lagrangian basin-scale ocean model used by Haertel and Fedorov [15] and the Lagrangian atmospheric model developed by Haertel et al. [4] This paper presents the first simulations conducted with the GLOM, which not only provide evidence of its usefulness as a climate modeling tool, but also help to illustrate the role of interior tracer mixing in maintaining global ocean circulation and stratification. This paper is organized as follows. Section 2 describes the components of the GLOM. Section 3 presents our first fully Lagrangian simulations of global oceans. Section 4 discusses the results presented in light of other studies.

2. Materials and Methods

The primary data source for this study is the Global Lagrangian Ocean Model (GLOM). It was developed by combining components of several previous Lagrangian ocean and atmosphere models. Each of these model components is briefly described in this section with the goal of providing the reader with an intuitive understanding of how they work. More complete technical details are available from the original references that discuss the development of particular model components.

2.1. Fluid Parcels

The GLOM represents a body of water as a collection of flexible water parcels [20]. Each parcel has a horizontal mass distribution function that remains fixed in the parcel's frame of reference (Figure 1).

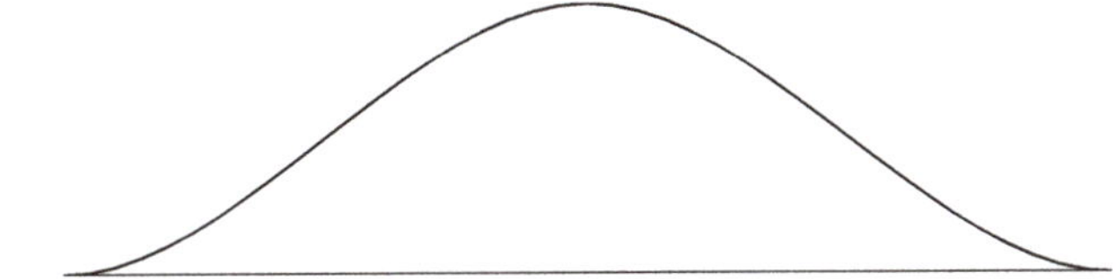

Figure 1. Parcel vertical thickness function (i.e., the outline of a parcel on a level surface).

The mass distribution function defines the amount of mass per unit of horizontal area between the top and bottom surface of the parcel. Since density variations in the ocean are small, this function also provides a good approximation of the shape of the vertical thickness of a parcel, and it is also referred to as the "vertical thickness" function. Here, we use a third-order polynomial in each horizontal dimension (latitude and longitude) to define the mass distribution function following [24]. Constructing the mass distribution function in this manner is computationally efficient, and it helps with conserving energy when calculating the pressure force, which also guarantees numerical stability [20]. Although this means that the horizontal projection of a parcel is a square, it turns out that most of the contours of vertical thickness are circular, as is illustrated in Figure 2b of [27].

Each column of mass within a given parcel can move up or down independently of neighboring columns as the parcel slides over variable bottom topography or other parcels. So, although all parcels have the same horizontal mass distribution or vertical thickness function (Figure 1), parcels have a variety of shapes owing to vertical shearing (e.g., Figure 2a). As parcels slide over one another, they conform, so there are no gaps between parcels. Since each parcel has a fixed amount of mass associated with it, parcel centers maintain an approximately uniform distribution. When two parcels meet, the parcel with a greater density slides under the parcel with a lesser density, so the fluid maintains a neutral or stable stratification (Figure 2a; note that darker shading denotes denser parcels).

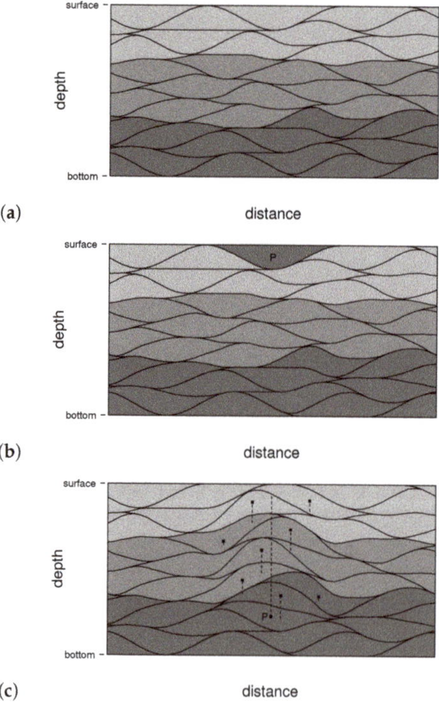

Figure 2. A schematic illustration of the Lagrangian convective parameterization. (**a**) A collection of water parcels, each having the same vertical thickness function, with darker shading denoting denser parcels. (**b**) The density of the top middle parcel (P) increases (e.g., from a temperature decrease or a salinity increase) yielding an unstable stratification. (**c**) The convective parameterization changes the stacking order of the parcels, so parcel P sinks to the level at which its density is the same as neighboring parcels. The long dashed line shows the path of the parcel center. Surrounding parcels rise slightly with short dashed lines illustrating the paths of parcel centers.

2.2. Lagrangian Convective Parameterization

On occasion, cooling or an increase in salinity can cause a near surface parcel to become more dense than parcels beneath it (Figure 2b). The Lagrangian convective parameterization then changes the stacking order of the parcels, so a neutral or stable stratification is restored (Figure 2c). Note that as the dense parcel sinks, neighboring parcels rise around it (Figure 2c; dashes lines denote vertical displacements of individual parcel centers). In this study, there is no mixing between convective updrafts and downdrafts (i.e., rising and sinking parcels). In atmospheric applications, we have found that carefully setting convective mixing (i.e., entrainment) is a key to accurately simulating moist convective systems such as the Madden Julian Oscillation [4–6].

Note that Figure 2 is a schematic illustration of the Lagrangian convective scheme, and it is intended to provide the reader with a mental picture, or intuitive understanding of how the parameterization works. For simplicity, the parcels have been drawn in two dimensions in an extremely coarse-resolution box-shaped ocean. A depiction of parcel interfaces for the simulations in this paper would include many more lines, a varying free surface elevation, and irregular bottom topography. Moreover, the GLOM does not actually keep track of parcel interfaces, but only the positions of parcel centers. The horizontal position vector for each parcel is a prognostic model variable (see below), whereas the vertical position of a given parcel depends on the vertical thicknesses of parcels beneath it. Near the beginning of each model time step, the vertical positions are diagnosed one parcel at a time starting with the lowest (most dense) parcel. The way that the convective scheme is implemented in the model is parcels are sorted by density after each time step, so density is a non-increasing function of the parcel array index, which determines the stacking order.

There are several limitations to this convective scheme. First, it is primarily intended to represent the net transports by deep, buoyancy-driven convection. The depth to which a parcel sinks is determined by its density relative to that of the environment (Figure 2). It is not intended to represent near-surface wind driven convection due to turbulent mixing, which can determine the depth of the surface boundary layer. Moreover, this scheme does not predict convective vertical velocities or accelerations of parcels, but rather the integrated effects of transports by updrafts and downdrafts. While mixing between rising and sinking parcels (i.e., entrainment) has been implemented and tested in atmospheric applications, it has not yet been tested in ocean simulations.

2.3. Model Equations

The following equations describe the motions of a fluid parcel:

$$\frac{d\mathbf{x}}{dt} = \mathbf{u} \tag{1}$$

$$\frac{d\mathbf{u}}{dt} + f\mathbf{k} \times \mathbf{u} = \mathbf{Ap} + \mathbf{Am} \tag{2}$$

where $\mathbf{x}$ is horizontal position, $\mathbf{u}$ is horizontal velocity, f is the Coriolis parameter, $\mathbf{Ap}$ is the acceleration due to pressure, and $\mathbf{Am}$ is the acceleration resulting from the mixing of momentum (i.e., viscosity). Evaluating the pressure acceleration involves approximating an integral of pressure over the surface of a parcel with a Riemann sum. Haertel et al. [24] developed a method to efficiently evaluate this sum for every parcel, and for technical details the reader is referred to that study. The pressure force is then divided by the parcel mass to yield the pressure acceleration vector. Vertical motion occurs when a parcel slides up or down variable topography, and/or when there is mass convergence or divergence beneath a parcel (i.e., as in conventional hydrostatic fluid models).

A fluid tracer q changes only in response to sources and mixing:

$$\frac{dq}{dt} = Sq + Mq \tag{3}$$

where Sq is the source of the tracer and Mq denotes mixing. In this study, the only tracer we consider is temperature, because we use the following simple equation of state:

$$\rho = 1029 \text{ kg m}^{-3} * (1 - 0.0002\, T) \quad (4)$$

where ρ is density, and T is temperature. The only temperature source for parcels is in the surface layer, where a restoring temperature is applied (see Section 2.7). The author chose not to include salinity's effects on density for several reasons. First, he judged the transition from basin-scale ocean simulations with idealized topography to global-scale simulations with realistic topography, along with the required rewrite of much of the model code, to be a task of sufficient complexity in and of itself without using a new equation of state. Second, a major goal of this paper is to extend the results of [15] to global scales and they did not use an equation of state with salinity. Third, the author's next planned application for the GLOM involves studying the impacts of air–sea coupling on the dynamics of the Madden Julian Oscillation, which does not require the use of an equation of state with salinity variations. In nature, of course, salinity variations make important contributions to density, so in interpreting the simulations presented here and comparing to observed ocean structure, it is probably best to think of temperature as a proxy for density.

2.4. Mixing

For the purposes of computing horizontal and vertical mixing, the domain is divided into isopycnal layers and columns respectively [24,25]. Columns of points are treated like columns of points in a finite difference model following [24]. Within isopycnal layers, parcels are allowed to mix momentum with their nearest neighbors following [25]. The Gent Mcwilliams [28] parameterization of mixing by eddies is not explicitly included, but Haertel and Fedorov [15] found that random motions of parcels generate a similar kind of mixing in this kind of Lagrangian model. The vertical viscosity is set to 10 $\text{cm}^2\ \text{s}^{-1}$, and the horizontal viscosity is set to $3 \times 10^4\ \text{m}^2\ \text{s}^{-1}$. For the simulation with interior mixing, the vertical tracer diffusivity is 1 $\text{cm}^2\ \text{s}^{-1}$ (it is zero in the simulation without interior mixing). The horizontal tracer diffusivity is set to zero for both simulations.

2.5. Bottom Topography and Spherical Geometry

The Lagrangian method requires a specification of the bottom surface elevation over a horizontal grid of points on which Riemann sums are evaluated to calculate the pressure force [24]. For the simulations presented in this paper, the surface elevation field is constructed in the following way. First, actual mean surface height of the earth's land surface and ocean bottom is averaged into 1-degree bins. Then, ocean depths are truncated at 4900 m, positive land surface values are set to 300 m, and the land mass of Central America is widened slightly. The resulting data are smoothed repeatedly using a 1-2-1 filter. Finally, the smoothed 1-degree data are averaged over the spherical grid of the ocean model. A comparison of the unsmoothed one-degree bathymetry (dashed) and the GLOM bottom topography (solid) is shown in Figure 3a for 30 W. Sharp topographic features are smoothed to be consistent with the large parcels used in this study, and the slope at the land ocean interface is reduced, which helps to improve numerical accuracy in low resolution simulations [25]. The land surface elevation is set to 300 m to prevent the smoothing of the bottom surface as well as the enhanced free surface height perturbations resulting from the use of gravity wave retardation [29] from overly distorting the locations of continental boundaries (e.g., from opening a water pathway through Central America).

The GLOM uses a spherical geometry in which grid boxes have a constant meridional width (in both degrees and meters), and a zonal width that is roughly the same in meters but increases in degrees longitude with distance from the equator. This geometry was first developed for a global Lagrangian atmospheric model by [4]. A mercator projection of the grid boxes is shown in Figure 3b for North America.

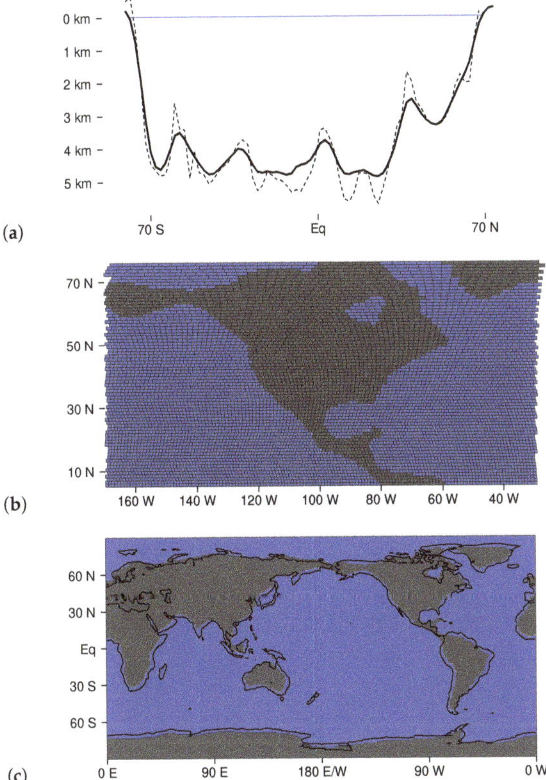

Figure 3. Bottom topography and spherical geometry. (**a**) Model bottom surface along 30 W (solid black line), the observed bathymetry (dashed line), and the undisturbed free surface height (blue). (**b**) Spherical geometry for North America. Each grid box used for calculating the pressure force is outlined in black. (**c**) Regions of land (gray) and ocean (blue) as determined by the zero height contour. The solid black line denotes the actual position of the land–sea interface.

Here, grid boxes are shaded gray for $z > 0$ m, and blue for $z < 0$ m. A similar shading of model bathymetry is shown for the entire world in Figure 3c, with the $z = 0$ contour for actual bathymetry delineated with a black line. Notice that small peninsulas and islands are largely smoothed out in the GLOM's bottom topography, but that the gross structure of continents and ocean basins is retained.

For the simulations discussed in this paper, grid boxes are 1 by 1 degrees wide at the equator, but have a larger width in degrees longitude at higher latitudes. Note that this grid is used to evaluate the pressure force and to create plots of layer thickness. In contrast, parcels span multiple grid boxes and move freely throughout the model domain. While it is difficult to precisely characterize the equivalent Eulerian resolution of a Lagrangian model, the author estimates it to be roughly 2 or 3 degrees at low latitudes, and somewhat coarser at high latitudes for the simulations presented in this paper.

2.6. Merging and Dividing Parcels

Following [25], the water in the oceans is divided into a collection of equally sized water mass elements (WMEs), which may be considered to be the building blocks of water parcels. The target size for parcels is one WME in the upper layer (0–700 m), two WMEs in the middle layer (700–2100 m), and four WMEs in the lower layer (depths greater than 2100 m). When a vertically thick parcel rises to a higher layer, and it is larger than the target size, it is sliced in half and the resulting two parcels begin moving independently (Figure 4).

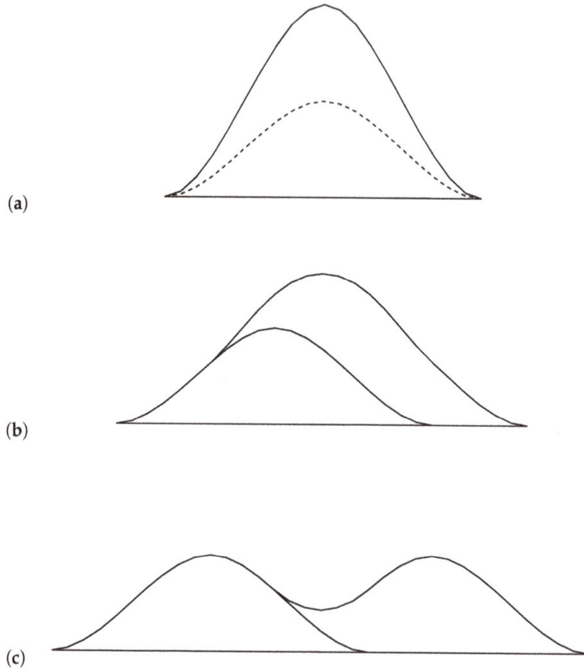

Figure 4. Dividing and merging parcels. (**a**) A single parcel comprising two water mass elements. (**b**,**c**) The parcel divides into two smaller parcels each comprising one water mass element, which then move independently. Viewing the panels in reverse order illustrates the merging process.

When a vertically thin parcel sinks to a lower layer, it maintains its size until it overlaps and is sufficiently close to another vertically thin parcel, at which time the two parcels are combined (Figure 4, view panels from the bottom up). Notice that each individual WME has its own identification number, and it can be tracked through the dividing and merging process. While merging generates a small amount of numerical mixing, Haertel and Fedorov [15] found little difference between simulations with dividing and merging when compared to simulations with vertically thin parcels everywhere. Presumably, this is because merging only occurs when parcels go through substantial vertical displacements (i.e., it is a rare event). While dividing and merging can be turned off to completely remove numerical mixing of tracers, the author selected to use it here because it dramatically speeds up the model (by a factor of 2 or more for the simulations presented in this paper). Owing to the dividing and merging process, the vast majority of parcels in the upper layer comprise a single water mass element and have a vertical thickness of approximately 78 m, with a factor of 2 (4) increase in vertical thickness in the middle (lower) layer. All parcels have a radius of 333 km in each horizontal dimension (i.e., 3 degrees latitude and 3 degrees longitude near the equator). There are approximately 154,000 water mass elements in each simulation, which allows the model to be run for millennial time scales on a single processor (the GLOM has not yet been coded to run in parallel) .

2.7. Experimental Design

In designing the simulations presented in this paper, the author had several goals: (1) to test the GLOM's ability to reproduce the gross circulation and stratification structure of the world ocean;

(2) to illustrate several unique capabilities of the GLOM that stem from its Lagrangian nature; and (3) to make a contribution to the field of physical oceanography. After reflecting on these goals, he chose to use the GLOM to extend the results of Haertel and Fedorov [15] (HF12) to include realistic topography and a global domain. Briefly, HF12 addressed the question of to what extent ocean circulation and stratification depend on interior mixing. They used a predecessor to the GLOM to simulate circulations in an idealized ocean with the scale of the Atlantic, and which included a circumpolar channel. They compared a simulation with a moderate amount of tracer mixing to one in which the tracer diffusivity was set to zero. They found that the leading order solution for ocean circulation, stratification, and heat transport could be reproduced with zero tracer diffusivity, and that interior mixing essentially contributed first-order perturbations to this solution. Accordingly, in this paper, we apply a surface forcing much like that used by HF12, but instead of using an idealized ocean basin, we use a global ocean with a smoothed version of actual bathymetry.

The temperature restoring function (Figure 5a, solid black line) is a piecewise linear function with a range set so that zonal average model SST (Figure 5a, red dashed line) has similar minima and maxima to that of zonal average SST in nature (Figure 5a, blue dotted line). Note that the restoring temperature is slightly lower in the Antarctic than in the Artic, which leads to Antarctic Bottom Water being more dense than North Atlantic Deep Water. The idealized zonal wind stress forcing (Figure 5b) is the same as that used by HF12; it includes strong westerlies at midlatitudes and weaker easterlies in the tropics.

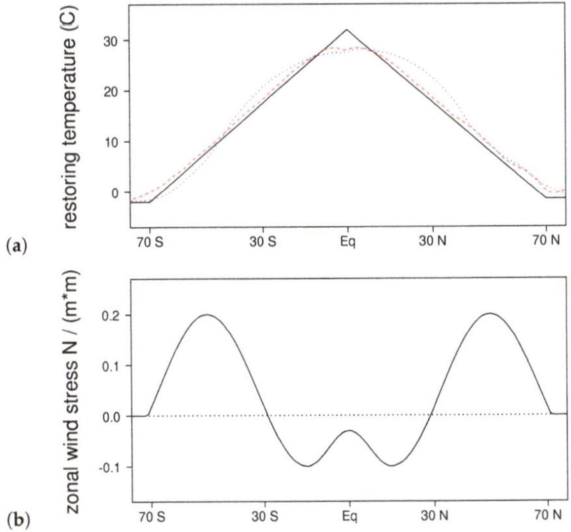

Figure 5. Surface forcing. (**a**) restoring temperature (solid black line), average SST in the simulation with mixing (red dotted line), and observed average SST (blue dotted line), which is from the NCEP Optimally Interpolated weekly SST and Sea Ice datasets for the years 1998–2009. (**b**) zonal wind stress.

The forcing was applied to an initially isothermal ocean (T = 0 °C) for a period of 1400 years. In one case, there was a moderately high vertical tracer diffusivity (1 $cm^2\ s^{-1}$), and in the other case the diffusivity was zero. One minor adjustment was made after 500 years of model integration–the temperature difference between the restoring temperature in the Artic and Antartic was reduced from 1 °C to 0.5 °C to deepen the Atlantic Meridional Overturning Circulation (AMOC). By the year 1400, the simulation with interior mixing was in an approximately steady state in the upper ocean, with very weak cooling in the abyss. In the simulation without interior mixing, weak cooling continued in the deep ocean and abyss. However, the temperature here had reached within a few tenths of a degree of the minimum restoring temperature, limiting the amplitude of possible future temperature changes.

3. Results

When the GLOM is forced with the temperature restoring and wind stress functions described in the previous section (Figure 5), it generates large-scale circulation patterns and stratification that are similar to those in the world ocean in nature. In this section, we examine these structures, and compare results for the simulation with interior mixing to the one without, as well as both simulations to observations. The results generally support the main conclusion of HF12—that a model with zero tracer diffusivity can produce most of the large-scale circulation and stratification structure seen in nature (i.e., the zero-order solution), with interior mixing contributing first-order perturbations.

3.1. Horizontal Stream Function

The horizontal circulation in the simulation with interior mixing (Figure 6a) is qualitatively consistent with that predicted by theory [30] for a forcing like that shown in Figure 5. Anticyclonic (cyclonic) gyres are present where the curl in the wind stress is negative (positive) with Sverdrup flow in the eastern portion of ocean basins, and more intense return flow in the form of western boundary currents to the east of continents. The model also produces an Antarctic Circumpolar Current (ACC), with gyres in the Ross and Weddell Seas. While the ACC is somewhat weaker than in nature [31], this is probably attributable to the low resolution of the model. In the simulation without mixing (Figure 6b), the overall flow structure is similar, but the amplitude of most gyres is slightly weaker, and the ACC is also weaker.

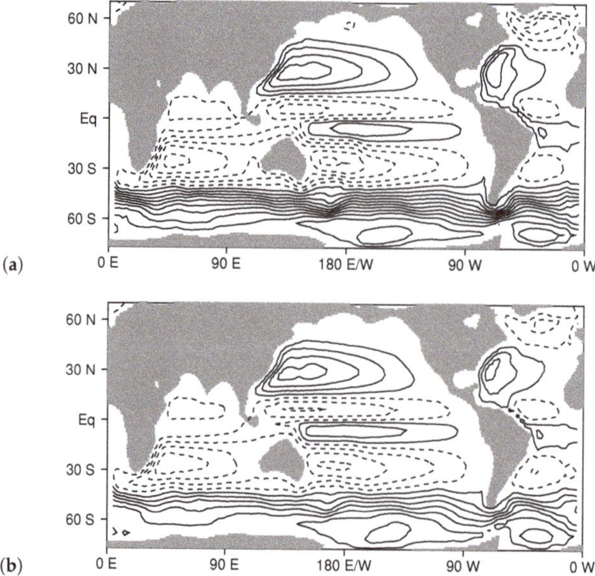

Figure 6. Horizontal streamfunction (10 Sv contour interval). (**a**) Simulation with interior mixing. (**b**) Simulation without interior mixing. Positive (negative) contours are drawn with solid (dashed) lines, with contour values ranging from −45 to 105 Sv.

3.2. Surface Temperature Field

Despite the idealized nature of the GLOM and the forcing, it qualitatively captures many of the departures from zonal symmetry seen in the observed sea surface temperature (SST) field, including warm tongues protruding poleward at mid-latitudes along the eastern boundaries of North America, Asia, Africa, and South America, an equatorial cold tongue in the eastern Pacific, and isotherms that slope northward from west to east across the North Atlantic (Figure 7).

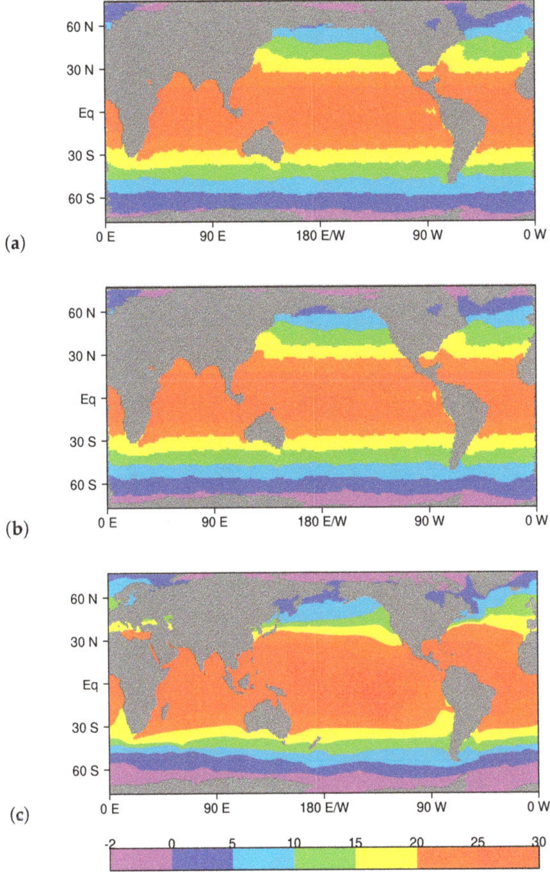

Figure 7. Sea surface temperature (°C). (**a**) Simulation with mixing. (**b**) Simulation without mixing. (**c**) Observed.

Of course, the low model resolution leads to western boundary currents that are broader and weaker than those observed in nature. Moreover, other features whose forcing is not included in the model are not well represented. For example, in nature, marine stratocumulus clouds are prevalent to the west of South America. They reflect a significant portion of the solar heating, reducing the SST there [32]. This forcing is not included in the GLOM simulations, and consequently SSTs are higher to the west of South America in the model (Figure 7a,b) than they are in nature (Figure 7c).

3.3. Stratification

In the simulation with interior mixing, the GLOM generates a region of water that is substantially warmer than that in the rest of the ocean, which is largely contained in the upper 1 km between 50° S and 50° N (Figure 8a).

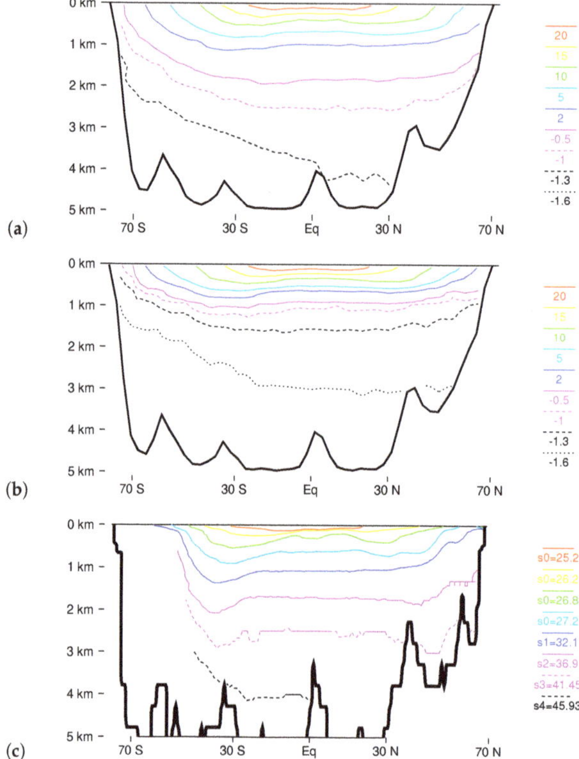

Figure 8. Stratification along 30° W. (**a**) Temperature (°C) in the simulation with mixing. (**b**) Temperature (°C) in the simulation without mixing. (**c**) Potential density in nature, with the upper four isopycals referenced to the surface, and the lower four isopycnals referenced to their approximate depths. The lower isopycnals are shown only in the vicinity of the depth to which they are referenced.

The vertical temperature gradient in the tropics within a few hundred meters of the surface is especially high. These aspects of the simulated thermocline structure are like those in the pycnocline in nature (Figure 8c). In both the model and in nature, the densest water forms in the Antarctic, and it undercuts the deep water formed in the north (Figure 8a,c). When interior tracer mixing is removed, the upper thermocline structure is very similar to that in the simulation with mixing (Figure 8b). However, at greater depths, the ocean warms substantially, with many isotherms rising around 1 km or more. The two simulations (with and without the mixing of tracers), have a similar slope in most of the Atlantic (Figure 8a,b), which is also similar to that of isopycnals in nature (Figure 8c).

3.4. Water Mass Distributions

One advantage of the GLOM is that it is easy to identify locations where water masses form. In Figure 9, we color code parcels along 30° W by the latitude at which they were last modified by the surface forcing (the model saves this information as a parcel variable and updates it it every time surface fluxes alter the parcel temperature). We compare the water masses in the GLOM (Figure 9a,b) to those in nature, as revealed by the salinity field (Figure 9c). In the simulation with mixing, each of the main water masses in the Atlantic is represented (Figure 9a): warm tropical and sub-tropical water near the surface that forms between 40° S and 40° N (red dots); Antarctic Intermediate Water (AIW) that forms between 40° and 60° S and moves northward along the main thermocline (cyan dots); North Atlantic Deep Water (NADW) that forms north of 40° N and reaches depths of about 4 km (green dots), and Antarctic Bottom Water (ABW) that forms south of 60° S, and spreads to cover most of the ocean bottom. Each of these water masses has a similar positioning to its counterpart in nature (Figure 9c), as inferred from the salinity field. In the simulation without tracer diffusion (Figure 9b), the general distribution of water masses is similar, except that the AIW penetrates farther north, the NADW is shallower, and the ABW is deeper.

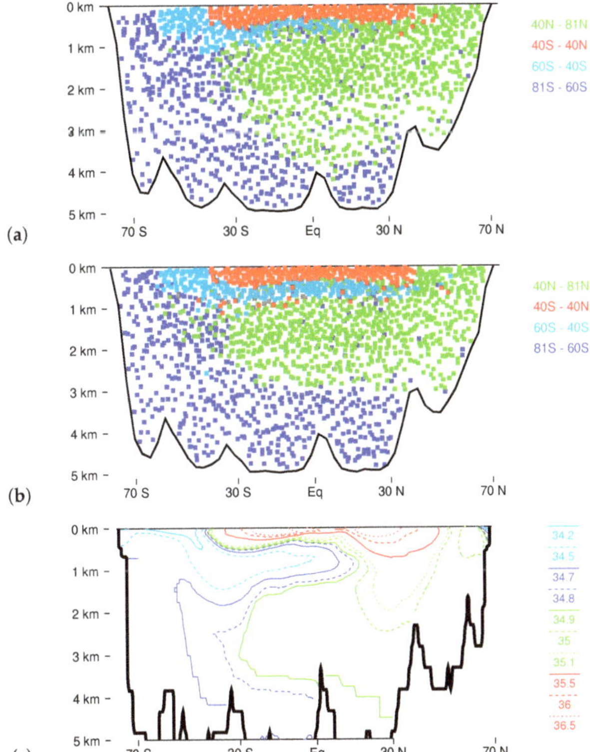

Figure 9. Water masses along 30° W. Parcels are color coded by the latitude of last surface contact in (**a**) the simulation with mixing. and (**b**) the simulation without mixing. (**c**) Observed water masses are inferred from the salinity field.

3.5. Atlantic Meridional Overturning Circulation

Most of the deep overturning the model generates is in the Atlantic Ocean, and we now examine this circulation in detail, both because of its importance to ocean heat transport and to follow up on the previous Lagrangian modeling results of HF12. When interior mixing is included in the model, the AMOC has a mid-latitude amplitude of about 21 Sv (Figure 10a), of which about 11 Sv upwells in the Southern Ocean. When interior tracer mixing is removed, the peak amplitude of the overturning is reduced to 13 Sv, but almost as much water (10 Sv) upwells in the Southern Ocean (Figure 10b). The overturning also becomes more shallow, with the deepest streamlines reaching around 3 km. The other difference is that there is a lack of streamlines crossing the thermocline between 27° S and 40° N. We also examine the AMOC using temperature as a vertical coordinate in Figure 11—both because this more directly illustrates heat transport by removing adiabatic overturning cells [15,33], and because it also more clearly illustrates shallow wind-driven cells. When interior mixing is included, the GLOM generates 24 Sv of deep overturning in the North Atlantic (Figure 11a), with roughly 12 Sv of water upwelling in the Southern Ocean. There is little temperature change following four streamlines (each representing 3 Sv of transport) from 70° N to 30° S. However, multiple streamlines indicate warming of water in the interior of the deep cell with roughly 6 Sv of upwelling near the equator (Figure 11a). When interior mixing is removed (Figure 11b), the amplitude of the deep cell is reduced to 15 Sv, but once again roughly 12 Sv of water sinks in the North Atlantic and makes a long journey to the Southern Ocean with very little temperature change. Also note that removing interior mixing has little impact on the shallow wind-driven cells for T > 10 °C. These results suggest that the basic character of the AMOC is maintained without interior mixing, but that mixing deepens the circulation, and generates transport across the equatorial thermocline.

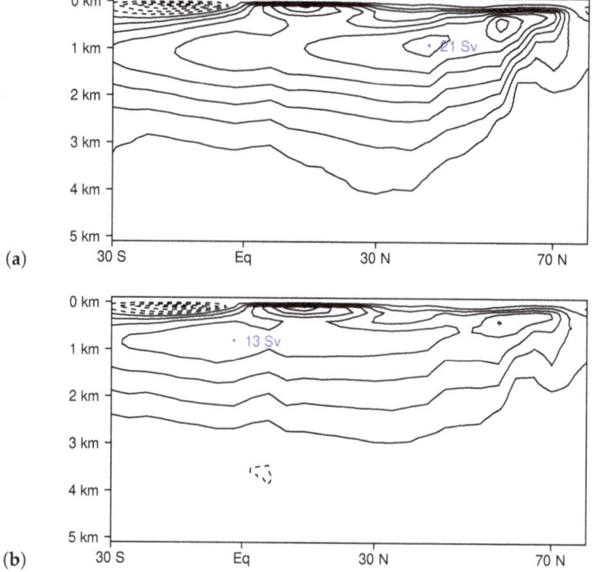

Figure 10. Meridional overturning streamfunction (3 Sv contour interval). (**a**) Simulation with mixing. (**b**) Simulation without mixing. Positive (negative) contours are drawn with solid (dashed) lines, with contour values ranging from −10.5 to 22.5 Sv.

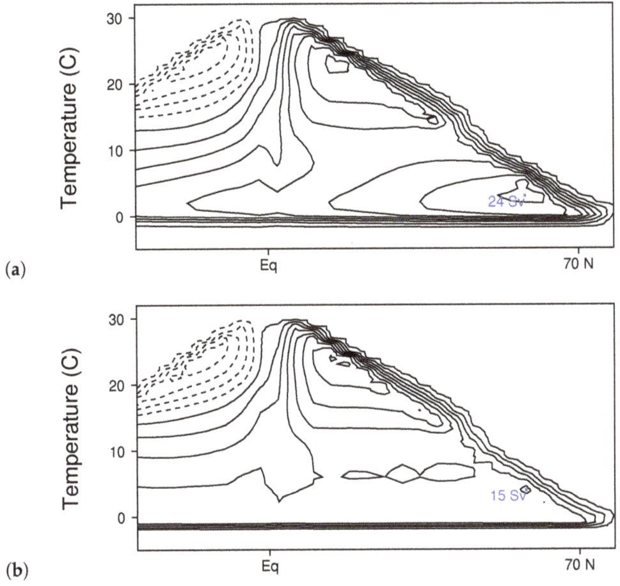

Figure 11. Meridional overturning streamfunction with temperature as a vertical coordinate (3 Sv contour interval). (**a**) Simulation with mixing. (**b**) Simulation without mixing. Positive (negative) contours are drawn with solid (dashed) lines, with contour values ranging from −10.5 to 22.5 Sv.

3.6. Sample Trajectory Analysis

One advantage of the GLOM is that it provides trajectory information for every wass mass element (WME) in the ocean. Each WME has a unique identification number (ID) that does not change during the course of a simulation. To compute a trajectory for a given WME, the modeler simply uses the ID to look up parcel positions for that WME for the times that data is saved. Moreover, for low-resolution runs, it is easy to construct trajectories for *every* WMI in the ocean. These can then be objectively partitioned to illustrate particular water pathways. We now perform such an analysis for the subsurface pathway of the AMOC.

Figure 12a shows downwelling (blue) and upwelling (red) locations of all WMEs that sink in the North Atlantic and upwell south of 30 N during the last 300 years of simulation with interior mixing. Water generally sinks (i.e., loses contact with the surface) in the Labrador or Norwegian Seas and upwells (i.e., regains contact with the surface) near the Gulf of Guinea or in the Southern Ocean. Regions most frequented by WMEs are contoured, revealing that the preferred pathway to upwelling is through a deep western boundary current just to the east of the Americas, which takes an eastward turn south of 20° S (see dashed and solid black contours in Figure 12a). When interior tracer mixing is removed, the downwelling locations are similar, as is the preferred pathway to the Southern Ocean, but water no longer upwells near the Equator (Figure 12b). Presumably, this is because without interior mixing, there is no longer any mechanism to warm the water once it loses contact with the surface, which would be necessary for equatorial upwelling.

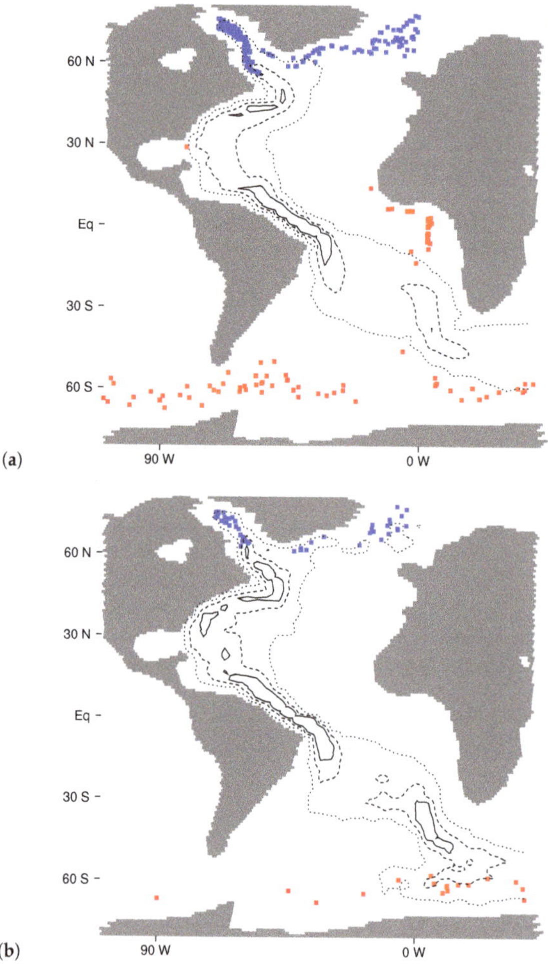

Figure 12. Horizontal pathway to upwelling of North Atlantic Deep Water. (**a**) Simulation with mixing. (**b**) Simulation without mixing. Blue (Red) dots indicated downwelling (upwelling) locations. Contouring indicates the percent of WME pathways that pass through each 3 by 3 degree grid box (10, 30, and 50 percent).

A y–z cross-section of the water pathways in shown in Figure 13.

In both simulations, water typically downwells just south of 70° N, and upwells near 60° S. In the simulation with interior tracer mixing, the pathway to upwelling is slightly deeper (Figure 13a), and includes a branch to equatorial upwelling that is not present in the simulation without interior mixing (Figure 13b). The pathways shown in Figures 12 and 13 are generally consistent with the streamfunctions shown in Figure 10; although the preferred pathway to upwelling is slightly shallower than the streamlines indicate, which may indicate a bias towards shallower paths because of the relatively short sampling time (300 years).

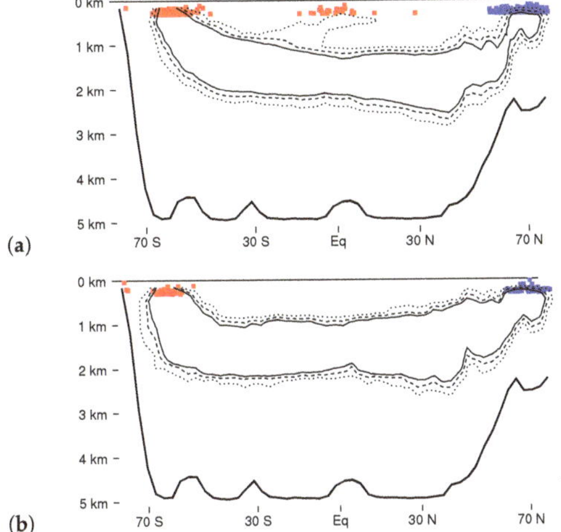

Figure 13. Vertical cross-section of pathway to upwelling of North Atlantic Deep Water. (**a**) Simulation with mixing. (**b**) Simulation without mixing. Blue (red) dots indicate downwelling (upwelling) locations. Contouring indicates the percent of WME pathways that pass through each 3 degree by 300 m grid box (10, 30, and 50 percent).

3.7. Pacific Water Masses

As in nature, the deep circulation in the GLOM in the Pacific Ocean differs greatly to how it is in the Atlantic Ocean. In particular, for both the simulations, with and without interior mixing, Antarctic Bottom Water fills almost all of the ocean at depths greater than 1 km (Figure 14). In the southern hemisphere, there are a few scattered parcels of North Atlantic Deep Water (green dots) at depths between 1 and 2 km that apparently come around the southern tip of Africa and through the Indian Ocean to reach the central Pacific. We conclude that there is essentially no deep water formation in the Pacific in the GLOM, which is not surprising considering the ocean geometry (Figure 3b) and the zonally symmetric temperature restoring (Figure 5b).

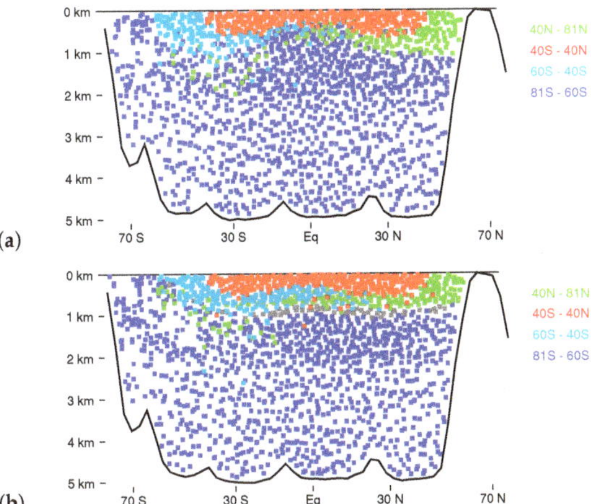

Figure 14. Water masses along 170° W. Parcels are color coded by the latitude of last surface contact in (**a**) the simulation with mixing, and (**b**) the simulation without mixing.

4. Discussion

In this paper, the author combines components of a Lagrangian basin-scale ocean model with the dynamical core of a global Lagrangian atmospheric model to create a new global Lagrangian ocean model (GLOM). Rather than numerically solving partial differential equations that describe fluid flow, this model predicts the motions of individual fluid parcels using ordinary differential equations and classical physics. The GLOM has variable bottom topography and spherical geometry, so it can be used to simulate general circulations of the global ocean. It also has a unique convective parameterization, in which the vertical positions of parcels are sorted by density to remove convective instability.

When forced with idealized surface temperature restoring and wind stress, the GLOM generates much of the circulation structure seen in nature in the global oceans, including mid-latitude gyres with western boundary currents, shallow wind-driven cells at low latitudes, a deep overturning in the Atlantic Ocean, and an Antarctic Circumpolar Current. The GLOM also produces thermocline structure and water mass distributions that are similar to those seen in nature. When interior mixing is removed, the large-scale ocean circulation, stratification, and water mass distributions are similar, with the main difference being that circulations, thermocline structure and upper-level water masses are slightly shallower. These results support the primary conclusion of [15] (and earlier papers such as [34,35]) that the leading-order solution for ocean stratification and circulation can be reproduced without interior tracer mixing, and that including tracer diffusivity creates first-order perturbations.

The work presented here and in previous Lagrangian modeling studies supports the idea that the Lagrangian approach would complement existing numerical methods used to study climate dynamics and climate change. This paper highlights several unique features of the GLOM: (1) the ability to conduct simulations with zero tracer diffusivity; (2) the unique convective parameterization which restacks parcels in convectively unstable regions, and (3) ease in tracking every mass element in the ocean and determining locations where water masses form. For these reasons, the author is continuing to refine and improve the GLOM, and he will soon be coupling it to a Lagrangian atmospheric model for coupled ocean–atmosphere climate dynamics experiments.

Of course, the GLOM has a number of disadvantages as well, so it would not be well suited for every climate application. For example, the model, as it is currently configured, uses a high degree of gravity wave retardartion (GWR) [29] which amounts to assuming that there is a layer of fluid above the ocean with a density slightly lower than that of salt water, and which slows externing gravity waves, allowing the model to have a large time step. The main side effect of GWR is that it greatly enhances the amplitude of free surface perturbations, which would be a problem for modeling circulations in shallow estuaries, for example. A second drawback to the Lagrangian approach used here is that there is a potential energy barrier to starting circulations in a pile of parcels [25], so it could have problems simulating weakly forced circulations.

The simulations presented in this paper are also limited in the sense that they use very large water parcels, with horizontal scales of a few degrees, and vertical scales on the order of 100 m. Owing to the lack of numerical tracer diffusion in the model, the equivalent resolution in an Eulerian ocean model is probably somewhat finer. For example, Haertel et al. [25] found that circulations and stratification in a 3-degree basin-scale Lagrangian ocean model compared favorably to those in a 1-degree z-coordinate model that was exposed to the same surface forcing. However, this resolution is still very coarse, and it represents the low end of expected climate applications (i.e., for millennial time-scale, single-processor, global simulations). Fortunately, there is reason to believe that the GLOM will have the capacity to be run at much finer resolution once it is coded in parallel. For example, Haertel et al. [24] found nearly linear scaling in a predecessor to the GLOM, which had similar computational costs to a sigma-coordinate ocean model for simulating circulations in a large lake. Moreover, it is encouraging that even at the very coarse resolution used in this paper, the GLOM was able to reproduce the gross circulation patterns and stratification seen in the world ocean, with the Lagrangian convective parameterization and random parcel motions apparently adequately representing buoyancy driven convective circulations and mesoscale eddy transports that occur at much smaller scales in nature.

Considering that the GLOM is in a relatively early stage of development when compared with other climate modeling tools, it is too early to fully understand its advantages and disadvantages at this time. Moreover, there are probably many other potential applications for the GLOM and its fully Lagrangian atmospheric counterpart that will take years to explore. For example, the recent study of Paparella and Popolizio [36] suggests that Lagrangian models will have advantages for simulating the mixing of biogeochemical tracers. However, one thing is clear already—Lagrangian models, as well as the physical parameterizations that go along with them, are fundamentally different from Eulerian models and methods, and they make a significant contribution to the diversity of climate modeling tools.

Funding: This research was supported by NSF grant AGS-1561066.

Conflicts of Interest: The author declares no conflict of interest.

References

1. Lin, J.; Brunner, D.; Gerbig, C.; Stohl, A.; Luhar, A.; Webley, P. *Lagrangian Modeling of the Atmosphere*; John Wiley & Sons: Hoboken, NJ, USA, 2013; Volume 200.
2. Bates, M.L.; Griffies, S.M.; England, M.H. A dynamic, embedded Lagrangian model for ocean climate models. Part I: Theory and implementation. *Ocean Model.* **2012**, *59*, 41–59. [CrossRef]
3. Van Sebille, E.; Griffies, S.M.; Albernathey, R.; Adams, T.P.; Befloff, P.; Biastoch, A.; Blanke, B.; Chassignet, E.P.; Cheng, Y.; Cotter, C.J.; et al. Lagrangian ocean analysis: Fundamentals and practices. *Ocean Model.* **2018**, *121*, 49–75. [CrossRef]
4. Haertel, P.; Boos, W.R.; Straub, K. Origins of Moist Air in Global Lagrangian Simulations of the Madden—Julian Oscillation. *Atmosphere* **2017**, *8*, 158. [CrossRef]
5. Haertel, P.; Straub, K.; Fedorov, A. Lagrangian overturning and the Madden–Julian Oscillation. *Q. J. R. Meteorol. Soc.* **2014**, *140*, 1344–1361. [CrossRef]
6. Haertel, P.; Straub, K.; Budsock, A. Transforming circumnavigating Kelvin waves that initiate and dissipate the Madden–Julian Oscillation. *Q. J. R. Meteorol. Soc.* **2015**, *141*, 1586–1602. [CrossRef]
7. Zhang, C. Madden-julian oscillation. *Rev. Geophys.* **2005**, *43*. [CrossRef]
8. Slingo, J.; Rowell, D.; Sperber, K.; Nortley, F. On the predictability of the interannual behaviour of the Madden-Julian Oscillation and its relationship with El Niño. *Q. J. R. Meteorol. Soc.* **1999**, *125*, 583–609. [CrossRef]
9. Lin, J.L.; Kiladis, G.N.; Mapes, B.E.; Weickmann, K.M.; Sperber, K.R.; Lin, W.; Wheeler, M.C.; Schubert, S.D.; Del Genio, A.; Donner, L.J.; et al. Tropical intraseasonal variability in 14 IPCC AR4 climate models. Part I: Convective signals. *J. Clim.* **2006**, *19*, 2665–2690. [CrossRef]
10. Hung, M.P.; Lin, J.L.; Wang, W.; Kim, D.; Shinoda, T.; Weaver, S.J. MJO and convectively coupled equatorial waves simulated by CMIP5 climate models. *J. Clim.* **2013**, *26*, 6185–6214. [CrossRef]
11. Maloney, E.D.; Hartmann, D.L. Modulation of eastern North Pacific hurricanes by the Madden–Julian oscillation. *J. Clim.* **2000**, *13*, 1451–1460. [CrossRef]
12. Fedorov, A.V.; Hu, S.; Lengaigne, M.; Guilyardi, E. The impact of westerly wind bursts and ocean initial state on the development, and diversity of El Niño events. *Clim. Dyn.* **2015**, *44*, 1381–1401. [CrossRef]
13. Wu, M.L.C.; Schubert, S.; Huang, N.E. The development of the South Asian summer monsoon and the intraseasonal oscillation. *J. Clim.* **1999**, *12*, 2054–2075. [CrossRef]
14. Neena, J.; Lee, J.Y.; Waliser, D.; Wang, B.; Jiang, X. Predictability of the Madden–Julian oscillation in the intraseasonal variability hindcast experiment (ISVHE). *J. Clim.* **2014**, *27*, 4531–4543. [CrossRef]
15. Haertel, P.; Fedorov, A. The ventilated ocean. *J. Phys. Oceanogr.* **2012**, *42*, 141–164. [CrossRef]
16. Griffies, S.M.; Pacanowski, R.C.; Hallberg, R.W. Spurious diapycnal mixing associated with advection in az-coordinate ocean model. *Mon. Weather Rev.* **2000**, *128*, 538–564. [CrossRef]
17. Bleck, R. An oceanic general circulation model framed in hybrid isopycnic-Cartesian coordinates. *Ocean Model.* **2002**, *4*, 55–88. [CrossRef]
18. Riehl, H. On the heat balance of the equatorial trough zone. *Geophysica* **1958**, *6*, 503–538.
19. Zipser, E.J. The role of organized unsaturated convective downdrafts in the structure and rapid decay of an equatorial disturbance. *J. Appl. Meteorol.* **1969**, *8*, 799–814. [CrossRef]

20. Haertel, P.T.; Randall, D.A. Could a pile of slippery sacks behave like an ocean? *Mon. Weather Rev.* **2002**, *130*, 2975–2988. [CrossRef]
21. Daly, B.; Harlow, F.; Welch, J.; Wilson, E.; Sanmann, E. *Numerical Fluid Dynamics Using the Particle-and-Force Method. Part I. the Method and Its Applications*; Technical Report; Los Alamos Scientific Lag., Univ. California: Oakland, CA, USA, 1964.
22. Monaghan, J.J. Smoothed particle hydrodynamics. *Annu. Rev. Astron. Astrophys.* **1992**, *30*, 543–574. [CrossRef]
23. Pavia, E.G.; Cushman-Roisin, B. Modeling of oceanic fronts using a particle method. *J. Geophys. Res. Oceans* **1988**, *93*, 3554–3562. [CrossRef]
24. Haertel, P.T.; Randall, D.A.; Jensen, T.G. Simulating upwelling in a large lake using slippery sacks. *Mon. Weather Rev.* **2004**, *132*, 66–77. [CrossRef]
25. Haertel, P.T.; Van Roekel, L.; Jensen, T.G. Constructing an idealized model of the North Atlantic Ocean using slippery sacks. *Ocean Model.* **2009**, *27*, 143–159. [CrossRef]
26. Van Roekel, L.P.; Ito, T.; Haertel, P.T.; Randall, D.A. Lagrangian analysis of the meridional overturning circulation in an idealized ocean basin. *J. Phys. Oceanogr.* **2009**, *39*, 2175–2193. [CrossRef]
27. Haertel, P. A Lagrangian method for simulating geophysical fluids. *Lagrangian Model. Atmos.* **2012**, 85–98. [CrossRef]
28. Gent, P.R.; Mcwilliams, J.C. Isopycnal mixing in ocean circulation models. *J. Phys. Oceanogr.* **1990**, *20*, 150–155. [CrossRef]
29. Jensen, T.G. Artificial retardation of barotropic waves in layered ocean models. *Mon. Weather Rev.* **1996**, *124*, 1272–1284. [CrossRef]
30. Pedlosky, J. *Ocean Circulation Theory*; Springer Science & Business Media: Berlin, Germany, 2013.
31. Cunningham, S.; Alderson, S.; King, B.; Brandon, M. Transport and variability of the Antarctic circumpolar current in drake passage. *J. Geophys. Res. Oceans* **2003**, *108*. [CrossRef]
32. De Szoeke, S.P.; Yuter, S.; Mechem, D.; Fairall, C.W.; Burleyson, C.D.; Zuidema, P. Observations of stratocumulus clouds and their effect on the eastern Pacific surface heat budget along 20 S. *J. Clim.* **2012**, *25*, 8542–8567. [CrossRef]
33. Döös, K.; Nycander, J.; Coward, A.C. Lagrangian decomposition of the Deacon Cell. *J. Geophys. Res. Oceans* **2008**, *113*. [CrossRef]
34. Luyten, J.; Pedlosky, J.; Stommel, H. The ventilated thermocline. *J. Phys. Oceanogr.* **1983**, *13*, 292–309. [CrossRef]
35. Welander, P. An advective model of the ocean thermocline. *Tellus* **1959**, *11*, 309–318. [CrossRef]
36. Paparella, F.; Popolizio, M. Lagrangian numerical methods for ocean biogeochemical simulations. *J. Comput. Phys.* **2018**, *360*, 229–246. [CrossRef]

Article

Seasonal Drought Forecasting for Latin America Using the ECMWF S4 Forecast System

Hugo Carrão [1,2], Gustavo Naumann [1,*], Emanuel Dutra [3], Christophe Lavaysse [1] and Paulo Barbosa [1]

1 European Commission, Joint Research Centre, 21027 Ispra, Italy; hugo.carrao@gmail.com (H.C.); christophe.lavaysse@ec.europa.eu (C.L.); paulo.barbosa@ec.europa.eu (P.B.)
2 Space4Environment, L-6947 Niederanven, Luxemburg
3 Instituto Dom Luiz, Faculdade de Ciências, Universidade de Lisboa, 1749-016 Lisbon, Portugal; endutra@fc.ul.pt
* Correspondence: gustavo.naumann@ec.europa.eu; Tel.: +39-0332-78-5535

Received: 4 May 2018; Accepted: 31 May 2018; Published: 1 June 2018

Abstract: Meaningful seasonal prediction of drought conditions is key information for end-users and water managers, particularly in Latin America where crop and livestock production are key for many regional economies. However, there are still not many studies of the feasibility of such a forecasts at continental level in the region. In this study, precipitation predictions from the European Centre for Medium Range Weather (ECMWF) seasonal forecast system S4 are combined with observed precipitation data to generate forecasts of the standardized precipitation index (SPI) for Latin America, and their skill is evaluated over the hindcast period 1981–2010. The value-added utility in using the ensemble S4 forecast to predict the SPI is identified by comparing the skill of its forecasts with a baseline skill based solely on their climatological characteristics. As expected, skill of the S4-generated SPI forecasts depends on the season, location, and the specific aggregation period considered (the 3- and 6-month SPI were evaluated). Added skill from the S4 for lead times equaling the SPI accumulation periods is primarily present in regions with high intra-annual precipitation variability, and is found mostly for the months at the end of the dry seasons for 3-month SPI, and half-yearly periods for 6-month SPI. The ECMWF forecast system behaves better than the climatology for clustered grid points in the North of South America, the Northeast of Argentina, Uruguay, southern Brazil and Mexico. The skillful regions are similar for the SPI3 and -6, but become reduced in extent for the severest SPI categories. Forecasting different magnitudes of meteorological drought intensity on a seasonal time scale still remains a challenge. However, the ECMWF S4 forecasting system does capture the occurrence of drought events for the aforementioned regions and seasons reasonably well. In the near term, the largest advances in the prediction of meteorological drought for Latin America are obtainable from improvements in near-real-time precipitation observations for the region. In the longer term, improvements in precipitation forecast skill from dynamical models, like the fifth generation of the ECMWF seasonal forecasting system, will be essential in this effort.

Keywords: drought; forecasting; Latin America

1. Introduction

Drought is a recurring and extreme climate event that originates in a temporary water deficit and may be related to a lack of precipitation, soil moisture, streamflow, or any combination of the three taking place at the same time [1]. Drought differs from other hazard types in several ways. First, unlike other geophysical hazards that occur along well defined areas (i.e., floods, earthquakes, landslides), drought can occur anywhere with the exception of desert regions and extremely cold areas where it does not have meaning [2,3]. Secondly, drought develops slowly, resulting from a prolonged period (from weeks to years) of precipitation that is below the average, or expected, value at a particular location [4].

To improve drought mitigation, different indicators are used to trigger a drought warning [1,5]. While an indicator is a derived variable for identifying and assessing different drought types, a trigger is a threshold value of the indicator used to determine the onset, intensity or end of a drought, as well as the timing to implement proper drought response actions [6,7]. Since precipitation is one of the most important inputs to a watershed system and provides a direct measurement of water supply conditions over different timescales, several commonly used drought indicators rely on precipitation measurements only [4]. Among them, the Standardized Precipitation Index (SPI) of [8] is certainly the most prominent; it has been recommended by the World Meteorological Organization (WMO) for characterizing the onset, end, duration and severity of drought events deriving from precipitation deficiencies taking place at different accumulation periods and occurring at different stages of a same hydro-meteorological anomaly [9].

The immediate consequences of short-term droughts (i.e., a few weeks duration) are, for example, a fall in crop production, poor pasture growth and a decline in fodder supplies from crop residues, whereas prolonged water shortages (e.g., of several months or years duration) may, among others, lead to a reduction in hydro-electrical production and an increase of forest fire occurrences [10]. Therefore, skillful predictions of the onset and end of a drought a few months in advance will benefit a variety of sectors by allowing sufficient lead time for drought mitigation efforts. Indeed, drought forecasting is nowadays a critical component of drought hydrology science, which plays a major role in drought risk management, preparedness and mitigation.

It has been demonstrated that droughts can be forecasted using stochastic or neural networks [11,12]. While [13] demonstrated that these type of forecast can provide "reasonably good agreement for forecasting with 1 to 2 months lead times", they do not quantify the improvement of these methods with respect to using probabilistic forecasts of the precipitation fields. Forecasts of droughts can also be produced using deterministic numerical weather prediction models. However, such forecasts are highly uncertain due to the chaotic nature of the atmosphere, which is particularly strong on a sub-seasonal timescale [14].

As an alternative, ensemble prediction systems that forecast multiple scenarios of future weather have considerably evolved over recent years. Indeed, the routine generation of global seasonal climate forecasts coupled with advances in near-real-time monitoring of the global climate has now allowed for testing the feasibility of generating global drought forecasts operationally. Systems to monitor drought around the globe are described in [7] for meteorological drought and in [15] for hydrologic and agricultural conditions. For example, Yuan et al. [16] used seasonal precipitation forecasts from the North American Multi-Model Ensemble (NMME) and other coupled ocean-land-atmosphere general circulation models (GCMs) to examine the predictability of drought onset around the globe based on the SPI. For the global domain, they found only a modest increase in the forecast probability of drought onset relative to baseline expectations when using the GCM forecasts. Hao et al. [17] described the Global Integrated Drought Monitoring and Prediction System (GIDMaPS) that uses three drought indicators. The forecasting component of their system relies on a statistical approach based on an ensemble streamflow prediction (ESP) methodology. Dutra et al. [18,19] generated global forecasts of 3-, 6-, and 12-month SPI by combining seasonal precipitation reforecasts from the European Centre for Medium-Range Weather Forecasts (ECMWF) System 4 (S4) with precipitation observations from the Global Precipitation Climatology Centre (GPCC) and, alternatively, the ECMWF Interim Reanalysis. They reported on several verification metrics for the SPI forecasts for 18 regions around the globe. Using the same definition as [16], they found that the ECMWF S4 provides useful skill in predicting drought onset in several regions, and the skill is largely derived from El Niño-Southern Oscillation (ENSO) teleconnections. However, they also found that in many regions is difficult to improve on "climatological" forecasts. Recently, Spennemann et al. [20] studied the performance and uncertainties of seasonal soil moisture and precipitation anomalies (SPI) forecasts over Southern South America by means of Climate Forecast System, version 2 (CFSv2). Their results show that both SPI and standardized soil moisture anomalies forecast skills are regionally and seasonally dependent. In general, a fast degradation of the forecast skill is observed as the lead time increases, resulting in almost no added value with regard to climatology at lead times longer than 3 months. However, they note that the forecasts have a higher skill for dry events if compared with wet events.

In this study, we build on the work of [18,19] by considering the ECMWF S4 ensemble framework to generate seasonal forecasts of the SPI, and perform their verification against corresponding SPI from precipitation observations of the GPCC over Latin America. Drought is viewed from a meteorological perspective, and seasonal forecasts of the 3- and 6-month SPI (SPI3 and SPI6) are generated and verified on a monthly basis for the hindcast period of 1981–2010.

While the focus of the work is on the prediction of meteorological drought, the study assesses two fundamental constraints in generating reliable regional drought predictions that will arise whether using the reported method or any other approach (e.g., land surface modeling): (1) the accuracy of summary statistics (e.g., mean, median, percentile) at predicting a seasonal drought from the members of the ensemble forecasting system; and (2) the skill of probabilistic categorical predictions of seasonal drought from the members of the ensemble forecasting system.

2. Study Area, Datasets and Methods

The study area covers the whole South-Central America region (the domain of analysis is limited to land surface grid points between 56° S–35° N, 33°–128° W). South-Central America spans a vast range of latitudes and has a wide variety of climates. It is characterized largely by humid and tropical conditions, but important areas have been extremely affected by meteorological droughts in the past [21–23] and the climate change scenarios foresee an increased frequency of these events for the region [24,25]. Given the significant reliance of South-Central American economies on rainfed agricultural yields (rainfed crops contribute more than 80% of the total crop production in South-Central America), and the exposure of agriculture to a variable climate, there is a large concern in the region about present and future climate and climate-related impacts [26]. South-Central American countries have an important percentage of their GDP in agriculture (10% average, [27]), and the region is a net exporter of food globally, accounting for 11% of the global value. According to the agricultural statistics supplied by the United Nations Food and Agriculture Organization [27], 65% of the world production of corn and more than 90% of the world production of soybeans occurs in Argentina, Brazil, the United States and China. The productivity of these crops is expected to decrease in the extensive plains located in middle and subtropical latitudes of South-America (e.g., Brazil and Argentina), leading to a reduction in the worldwide productivity of cattle farming and having adverse consequences for global food security [28,29].

2.1. Forecasts: The ECMWF Seasonal Forecast System (S4)

In this study, we use the ECMWF seasonal forecast system 4 (hereafter S4; [30]) to forecast 3- and 6-month SPI. The S4 is a dynamical forecast system based on an atmospheric-ocean coupled model, which has been operational at ECMWF since 2011 and is launched once a month (on the first day of the month). The 2011 version of the forecast model has 91 vertical levels, lead times up to 13 months, and a resolution of T255 (80 km). It provides back integrations (hindcast) with 15/51 member ensemble (number depends on month) for every month from 1980 onwards. Molteni et al. [30] provide a detailed overview of S4 performance. For the comparison with the GPCC observations, the S4 has been re-gridded to 1.0° latitude/longitude grid spacing, and daily precipitation values over its hindcast period (1981–2010) have been aggregated to monthly values. The ability of the probabilistic model to accurately forecast seasonal drought conditions has been evaluated up to 6 months of lead time. In addition to the dynamical seasonal forecasts and in order to test whether the forecasts perform better than a benchmark, a set of climatological forecasts (CLM) were also generated by randomly sampling past years from the reference data set to match the number of ensemble members in the hindcast, as depicted in [19].

2.2. Observations: The GPCC Full Data Reanalysis Version 6.0

In this study, monthly precipitation totals at 1.0° latitude/longitude grid spacing from the Full Data Reanalysis Monthly Product Version 6.0 of the GPCC are used as a reference data set (for the forecast verification). The GPCC was established in 1989 on request of the World Meteorological Organization (WMO) and provides a global gridded analysis of monthly precipitation over land from operational in situ rain-gauges based on the Global Telecommunications System (GTS) and historic precipitation data measured at global stations. The data supplies from 190 worldwide national weather services to the GPCC are regarded as primary data source, comprising observed monthly totals from 10,700 to more than 47,000 stations since 1901. The monthly gridded data sets are spatially interpolated with a spherical adaptation of the robust Shepard's empirical weighting method [31]. Validation of the original data sets for drought monitoring has been performed by [18,32], who found that GPCC data sets show higher values for extreme precipitation, and tend to over-smooth the data. This can generate some problems when analyzing intense precipitation events but appears of secondary importance in drought analysis. Therefore, to be consistent with the data provided by the ensembles from ECMWF, a common period of the hindcast that covers the period from 1981 to 2010 is used to calculate the SPI.

2.3. Drought Indicator: The Standardized Precipitation Index (SPI)

In this study, we selected the SPI [8] as a meteorological drought indicator. The SPI is a statistical indicator that compares the total precipitation received at a particular location during a period of time with the long-term precipitation distribution for the same period of time at that location. In order to allow for the statistical comparison of wetter and drier climates, the SPI is based on a transformation of the accumulated precipitation into a standard normal variable with zero mean and variance equal to one. SPI results are given in units of standard deviation from the long-term mean of the standardized precipitation distribution. Negative values, therefore, correspond to drier periods than normal and positive values correspond to wetter periods than normal. The fundamental strength of the SPI is that it can be calculated for a variety of precipitation timescales (e.g., weekly, monthly, seasonal or yearly accumulation periods) and updated on various time steps (e.g., daily, weekly, monthly), enabling water supply anomalies relevant to a range of end users to be readily identified and monitored. SPI is typically calculated on a monthly basis for a moving window of n months, where n indicates the precipitation accumulation period.

The magnitude of negative SPI values correspond to percentiles of a probability distribution that are frequently used as threshold levels (triggers) to classify drought intensity [8,33,34]. Several classification systems of meteorological drought intensity based on fixed threshold levels of the SPI have been presented in the literature. The most widely known is that proposed by [8], which maps precipitation totals below the 50th percentile into four fixed categories of drought intensity (Table A1). For example, a "moderate" drought event starts at SPI = -1.0 (units of standard deviation), which corresponds to a cumulative probability of 15.9%, that is, approximately the 16th percentile. McKee at al. [8] determined that every region is in "mild" drought 34% of the time, in "moderate" drought 9.2% of the time, in "severe" drought 4.4% of the time, and in "extreme" drought 2.3% of the time (Table A1). The threshold levels of drought intensity proposed by [8] have been used worldwide in numerous applications at different timescales of precipitation accumulation, such as to monitor drought in the United States [35,36] and Europe [37], for detecting droughts in East Africa [38], to monitor drought conditions and their uncertainty in Africa using data from the Tropical Rainfall Measuring Mission (TRMM) [32], and for improving the fire danger forecast in the Iberian Peninsula [39].

2.4. Drought Detection and Verification Methods

The methods to detect drought events from the S4 ensemble system (Table A2) were defined in [40] as 13th percentile (Q13); 23th percentile (Q23); Median (MED); 77th percentile (Q77); 88th percentile (Q88); Large spread (SpL); Low spread (Spl); Dry spread (SpD); Flood spread (SpF); Mean (EM_RES).

Forecast verification is the process of assessing the quality of forecasts. The usefulness of forecasts to support decision making clearly depends on their error characteristics, which are elucidated through forecast verification methods. In this study, the forecasts correspond to the monthly SPI3 and SPI6 values computed with the ECMWF S4 for the period 1987–2010; the observations correspond to the SPI3 and SPI6 values computed with the GPCC for the same historical period. The validation methods used are the percentage correct (PC), extreme dependency score (EDS), Gilbert skill score (GSS), BIAS, probability of detection (POD), and False Alarm Rate (FAR). A comprehensive description of the validation metrics can be found in the supplementary material.

3. Results and Discussion

Initially, we assessed the ability of the ECMWF S4 ensemble system to seasonally forecast the spatial distribution of SPI in South-Central America by evaluating its monthly scalar accuracy and skill score at each location with 3- and 6-month lead time (respectively for the SPI3 and SPI6). In the sequence, we verify the non-probabilistic identification of drought events by means of the S4 system.

3.1. Non-Probabilistic Forecasts of Continuous SPI Values

In Figure 1, we present the monthly correlation between observed and forecast ensemble mean (a) SPI3 and (b) SPI6 at, respectively, 3- and 6-month lead time for the hindcast period of 1981–2010. The maps depicted in Figure 1 show that there is a positive correlation between SPI3 forecast and observations at all months and for most of the study area. Overall, the forecast SPI3 values follow the trends (increases or decreases) of the observed SPI3 values. Notwithstanding, the statistical significance between observed and forecast SPI3 varies across regions and months: for example, the correlation along the East Pacific coast is almost never statistical significant during the year, it is mostly statistical significant during the whole year for Northeast of South-America, and significant patterns are only verified for Central America during the months between December and May. On the other hand, SPI6 forecasts present extensive geographic areas that are negatively correlated with SPI6 observations at 6-month lead time (Figure A1). These large forecast errors are not systematic but occur mainly for the Amazon and Central East part of South America, and are most evident during the months of January–April (end of the wet season) and June–August (dry season). Surprisingly, and similarly to the SPI3, the correlation is statistically significant during almost the whole year for the Northeast of South America and for large parts of Central America from March to May. Mo and Lyon [41] suggest that the statistically significant correlation patterns in Central America and Northeast of South America are likely contributed by the ENSO: these regions are known to have a strong ENSO signal, and the seasonal skillful of precipitation forecasts contribute to the SPI3 and SPI6 seasonal forecasts. Moreover, in those areas and during both seasons (wet and dry), the intra-seasonal patterns of precipitation seem to be highly influenced by the activity of the Madden–Julian Oscillation [42]. Since the correlation is statistical significant for some regions at some months, then it suggests that the forecast has some skill at 3- and 6-months lead time.

The scalar skill score was also analyzed to assess the ability of the forecasts to improve SPI prediction over the climatological median values (i.e., SPI = 0). The differences between the ECMWF-based forecasts and the climatological forecasts (CLM) will indicate whether there is additional skill obtained from the dynamical model forecasts. In Figure 2, we present the monthly SPI3 forecast skill (using the mean of the ensemble) at 3-month lead time relative to baseline skill for the hindcast period 1981–2010, which shows the difference in correlation between the ECMWF S4 SPI3 forecasts and the baseline SPI3 forecasts based on climatological probabilities. Our results confirm that the forecasts have higher skill than the baseline, but the differences are often not significant at the 5% level based on the Fishers Z test. Indeed, although the correlation with observations is extensively significant over the study area, it does not extensively improve over the climatological SPI values. Marked improvements are observed for Northeast Brazil during the months of April–July, Mexico during the months of December–April, and North of South America between January–April. Overall, our results are consistent with [19,41], namely, that it is still challenging to improve on SPI forecasts that are based on climatology and persistence.

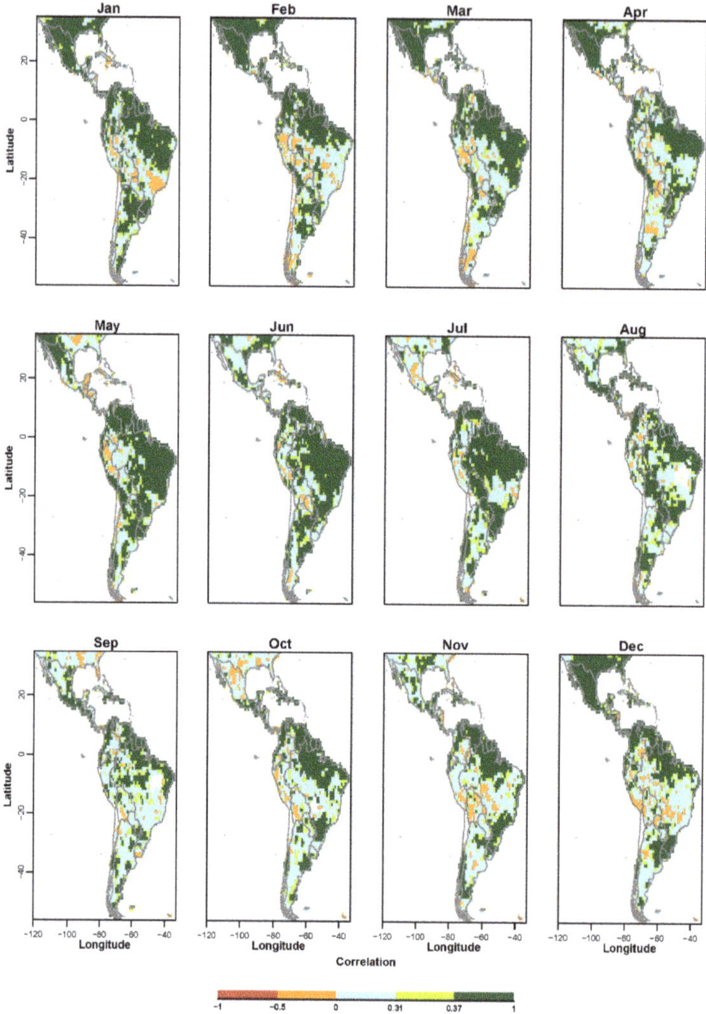

Figure 1. Monthly correlation of the observed and forecast standardized precipitation index (SPI) at 3-months lead time (SPI3) (using the mean of the ensemble) for the hindcast period (1981–2010). Values are indicated in the color bar: 0.31 (0.37) is statistical significant at 10% (5%) significance level.

Interestingly, scalar skill score results suggest that SPI3 forecasts match the observations in dry regions mainly during the beginning of the dry seasons, while at regions with high rainfall variability and/or during the wet seasons the forecasts are usually less skillful. Therefore, we believe that the ECMWF S4 ensemble mean might underestimate monthly rainfall and thus increase the intensity of dry periods and lessen the forecast values of SPI3 for the study region.

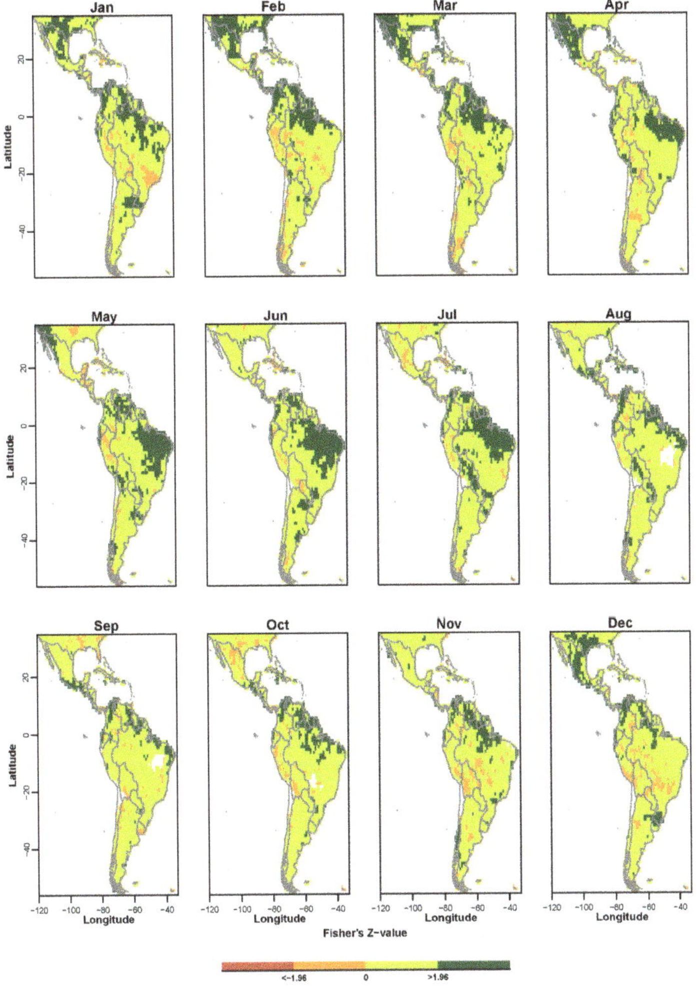

Figure 2. Monthly difference in forecast skill (Pearson correlation) between the forecast SPI3 at 3-month lead time (using the mean of the ensemble) and climatological SPI for the hindcast period (1981–2010). Values are indicated in the color bar: 1.96 is the statistical significance at the 5% significance level.

On the other hand, the 6-month seasonal forecasts are less skillful than the 3-month forecasts (Figure A2). Indeed, and as expected from the correlation analysis, skill scores for the SPI6 forecasts are generally lower than for SPI3 and almost not statistically significant at the 5% level. In Figure A2, it is perceptible that regions with meaningful SPI6 forecasts are also depicted as skillful for the SPI3. The monthly skill scores clearly show that the meaningful forecasts are concentrated over the eastern Amazon, namely in most of the states of AP (Amapá), PA (Pará) and MA (Maranhão). Molteni et al. [30] states that some important bias reductions were introduced in S4, as compared to S3, particularly in the tropical Atlantic and Indian Oceans, and some improvements over land areas e.g., in East Asia and over the Amazon Basin. It is possible that these improvements over the bias of the ECMWF S4 precipitation forecasts will reduce the residual errors between observed and predicted seasonal SPI values.

In Figure 3, the Root Mean Squared Error (RMSE) values between observed and forecast SPI3 at 3-month lead time (Figure A3 for SPI6), for the hindcast period 1981–2010. The results suggest that the predicted SPI is less consistent with the observations derived from GPCC for those regions placed in the subtropical subsidence zones around 10° and 30° N/S, such as subtropical southeast and central Brazil, Paraguay and Bolivia, as well as large areas of Peru.

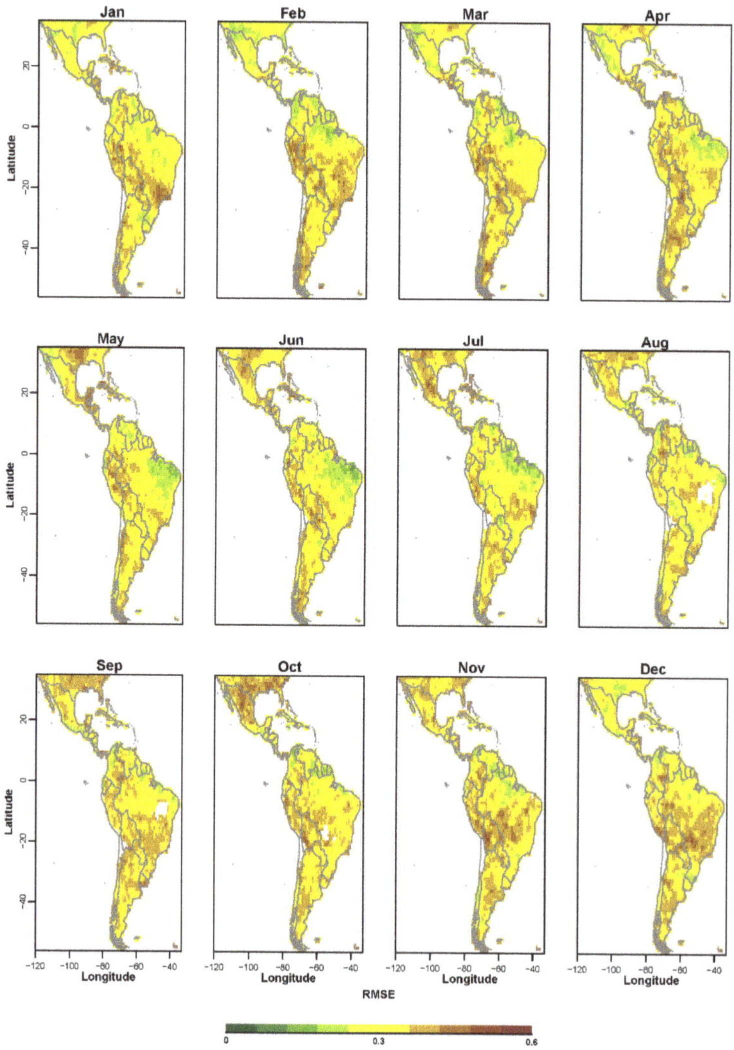

Figure 3. Root Mean Squared Error (RMSE) between the observed and forecast SPI3 at 3-month lead time (mean of the ensemble) for the hindcast period (1981–2010). Values in difference of percentile magnitude are indicated in the color bar.

The high variability of precipitation regimes within those latitudes [43,44] makes it difficult to predict drought at seasonal scale. The results based on the analysis of residual errors also suggest that locations with monthly forecast errors inferior to 0.2 have significant skill, whereas those superior to 0.5 have negative correlation and are unskillful. This output is confirmed by the monthly skill score measured in terms of the RMSE (Figure 4). The RMSE skill score approximates the skill score computed with the correlation index (Figure 2) and its spatial patterns: overall, seasonal SPI3 and SPI6 forecasts are monthly skillful for a small region in the eastern part of the Amazon Basin.

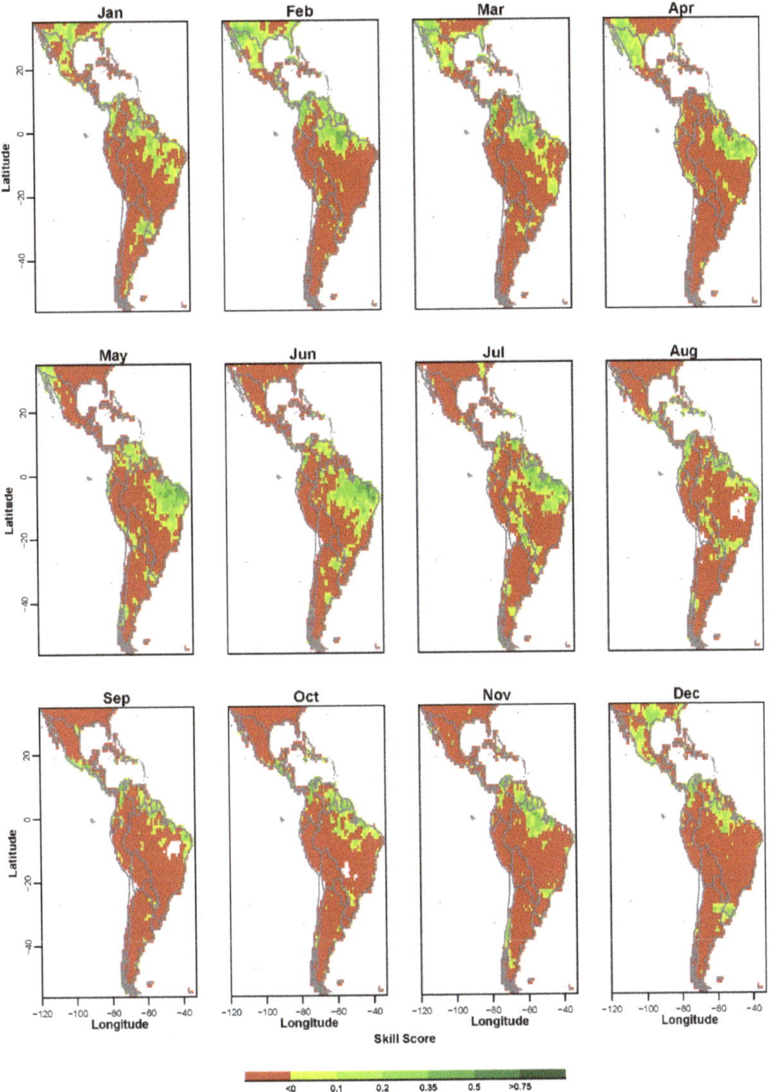

Figure 4. Skill Score of the SPI3 at 3-month lead time forecast measured in terms of the RMSE relative to climatological RMSE for the hindcast period (1981–2010).

3.2. Non-Probabilistic Forecasts of Categorical SPI Values

In Figures 5 and A5 the score values of categorical drought forecasts are represented (i.e., below the SPI -1 threshold) while the ensemble drought detection was based on several methods as depicted in Table A2. We have pooled together all seasons and locations at the study area in generating Figures 5 and A5.

Surprisingly, the distribution of score values for SPI3 and SPI6 are alike for all methods and all verification measures. This may be due to the fact that boundary conditions of seasonal dynamical model forecasts are often characterized by low frequency variability, leading to similar predictability of medium-range climate conditions that extend from a few to several months lead time. In general, precipitation is the result of a complex and interacting phenomena at different spatial and temporal scales, but regional atmospheric patterns that are actively involved in the development of long-term drought conditions are persistent and influenced by predictors that can be accurately estimated at large lead times. Therefore, precipitation anomalies over extreme peak thresholds (drought conditions) might be similarly predicted for different accumulation periods and seasonal lead times, although the accuracy of their scalar values may vary regionally and seasonally. Moreover, given the similar distribution of score values for different methods of categorical drought identification, we present the results of the SPI3 and SPI6 in a joint analysis.

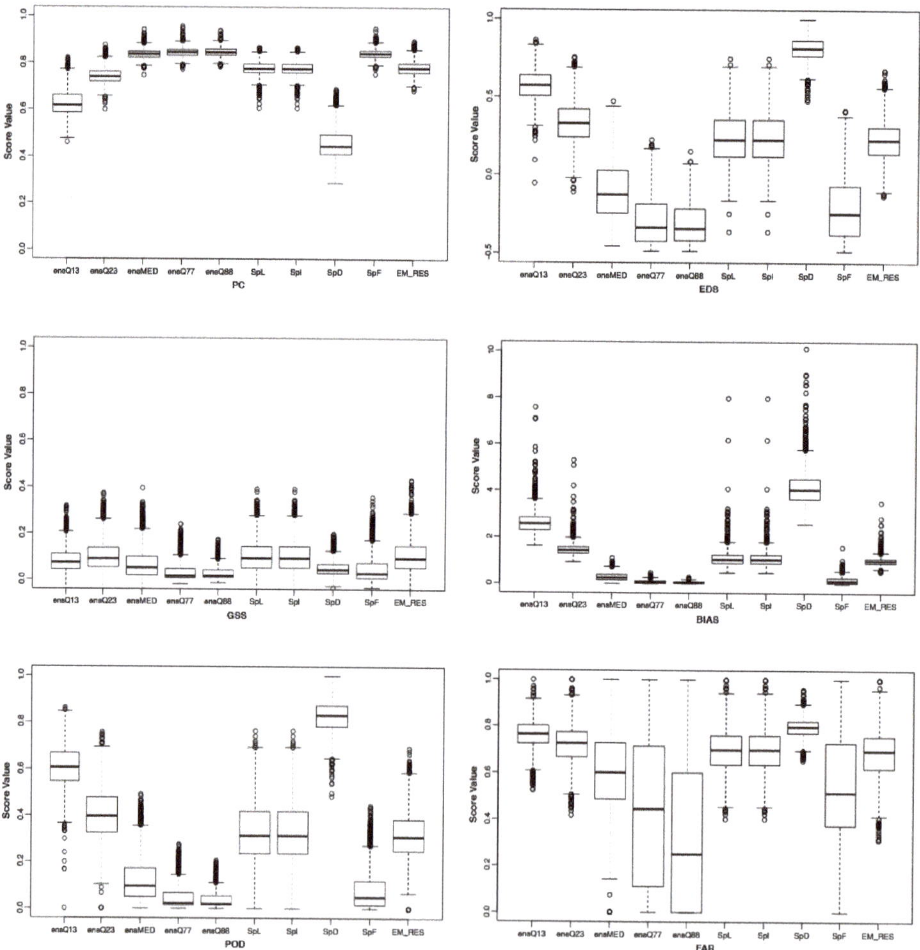

Figure 5. Verification measures of categorical drought forecasts (i.e., below the SPI3 "-1" threshold) estimated with the methods described in Table A2.

For categorical drought events predicted with both SPI3 and SPI6, computed with the ECMWF S4 ensemble mean (EM-RES), POD values indicate that for at least 50% of the locations in South-Central

America one in three seasonal drought events is correctly predicted. This is better than the respective climatology (16% of drought events are correctly detected) and extends over a geographic area larger than that with statistical significant scalar skill scores. Although the ensemble mean performs better than the climatology, POD values are still higher for the methods Q13 (60% of detection) and SpD (80% of detection); the worst results of all the methods are given by the wettest members of the ranked distributions (Q77 and Q88). This means that drier members are better than the mean at detecting the drought onset, but also that there is a low consistency between the extreme and dry members of the ECMWF S4 ensemble set. Lavaysse et al. [40] found similar results in in Europe, where the highest POD is achieved by using the 13 percentile, and the product using the Q13 and Q23 (SpD).

According to the FAR scores, we perceive that by using the ensemble mean SPI values to correctly detect a drought (EM_RES), there will be on average a 70% rate of false alarms. Median FAR values are even larger for dryer members (10% more for Q13 and SpD), and the inter-quantile range of the wettest members is about six times greater than that of the mean (60%), which indicates a large spread of FAR values. Based on these results, it is difficult to select the method that better optimizes between the number of drought hits and the number of misses. Indeed, while the mean of the ensemble shows always an average number of hits and misses (as similar to Spl and SpL, which represent the mean of ensemble extreme and opposite members), the dryer and wetter members of the ensemble attain, respectively, extreme numbers of hits or misses. In that sense, Lavaysse et al. [40] proposed a way to compensate the effect of number of event detected in POD and FAR by using specific thresholds in order to select the same number of events for the different methods.

Looking at PC, we might suggest that the Q13 of the ensemble is the worst performing method to detect between drought and non-drought events. On the other hand, by looking the EDS we might suggest that Q13 is the best method to detect the onset and end of a drought. Because of the non-dependency of the EDS alone to assess a model's performance on size is fixed, Ghelli and Primo [45] have suggested to not use the EDS alone to assess a model's performance on forecasting rare events. Those authors have shown that the EDS equation results in an increased freedom of false alarms and correct negatives, which can freely vary with the only restriction that their sum has to be constant. This feature encourages hedging, that is, forecasting the event all the time to guarantee a hit and thus to ensure a higher success rate, however this will increase the false alarm ratio and bias. Therefore, it is paramount to use the EDS in combination with other scores that include the right hand side of the contingency table, as the false alarm rate and/or the bias. Indeed, both FAR and BIAS show that SpD is not an accurate method to detect drought, as it forecasts a large number of drought events that do not occur.

In that sense [40] proposes the use of the maximum Gilbert skill score (GSS) as trigger-point to find the method that better optimizes among the number of false alarms, misses and hits of drought events identified with the SPI. Looking at Figures 5 and A5, it is noted that the ensemble mean (EM_RES) is the best choice for discriminating among seasonal drought and non-droughts events at 3- and 6-month lead time, whilst keeping a minor number of false alarms. Although the SpD gives the best POD, it also increases the ratio of false alarms and diminishes the overall skill score of the method. Following the approach by [40], we suggest that the ensemble mean should be used to trigger the warning of seasonal drought events for South-Central America by means of the SPI3 and SPI6 for respectively 3- and 6-month lead times.

3.3. Probabilistic Forecasts of Categorical SPI Values

In addition to having skillful forecasts of scalar SPI3 and SPI6 derived with the ECMWF S4 ensemble mean at seasonal lead times, a second fundamental challenge to generate reliable drought forecasts for the region is associated with uncertainties in the ensemble used. Therefore, to further quantify the uncertainties arising from the spread of the ensemble when computing the SPI, we computed the overall Brier Skill Score (BSS), based on the climatological frequency of "moderate", "severe" and "extreme" drought events (Table A1). In Figures 6 and 7, we map the spatial distribution of BSS for the ECMWF S4 SPI-3 and SPI6 forecast respectively, measured in terms of the BS relative to climatological BS at a lead time of 3 and 6 months for the hindcast period 1981–2010. We have pooled together all seasons at each grid point.

The spatial distribution of BSS suggest that the skill of the forecasting system is very similar for both accumulation periods and decreases with the increasing intensity of drought. Looking at the skill for predicting "moderate" drought events, the maps introduced in Figures 6 and 7 show that the forecasting system behaves better than the climatology for large clustered points at the North of South America, Mexico, Northeast of Argentina and Uruguay. In the later regions, where a hot spot appears over La Plata basin, local feedbacks between soil properties and precipitation variability can explain the improved skill which is linked to the coupling strength between soil moisture, evapotranspiration, and temperature [46,47]. On the other hand, the system skill for predicting "extreme" drought events is limited to a few locations in Northeast Brazil, Northeast Mexico, Northeast Amazon, and Northeast of Argentina. These results are encouraging, but only the Northeast of Mexico shows some spatial clustering with positive BSS for extreme drought events, while positive BSS is spatially scattered for the other regions. On combining these results, it can thus be reasonably assumed that forecasting different magnitudes of meteorological drought intensity on seasonal time scales remains quite challenging, but the ECMWF S4 forecasting system does at least a promising job in capturing the drought events (i.e., "moderate" drought) for some regions.

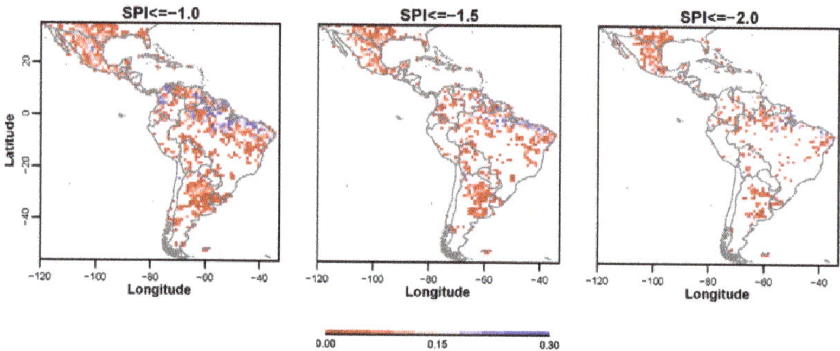

Figure 6. Brier Skill Score (BSS) of the European Centre for Medium Range Weather (ECMWF) S4 SPI-3 forecast for different probabilities of SPI occurrence, at a lead time of 3 months for the hindcast period 1981–2010. Values are indicated in the color bar; land grid points colored in white indicate that the forecasting system is no more skillful than the climatology.

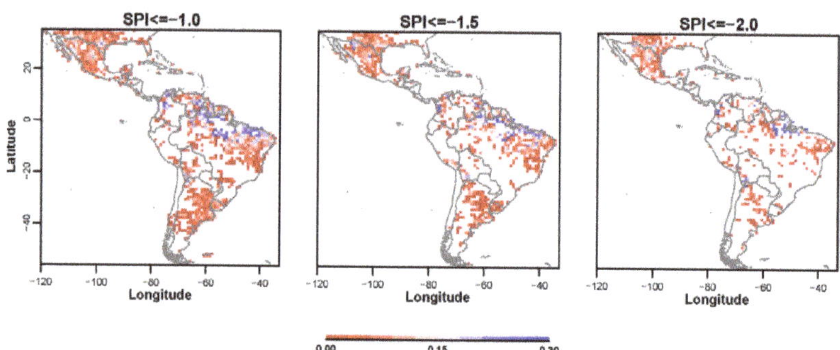

Figure 7. Brier Skill Score (BSS) of the ECMWF S4 SPI-6 forecast for different probabilities of SPI occurrence, at a lead time of 3 months for the hindcast period 1981–2010. Values are indicated in the color bar; land grid points colored in white indicate that the forecasting system is no more skillful than the climatology.

It is interesting to note that the spatial pattern of positive BSS at different SPI categories closely matches the regions that show significant skill scores for non-probabilistic drought forecasts, as well as the geographic grid points that have the lowest monthly RMSEs (Figure 3). As expected, the BSS is lower for the locations where the scalar mismatch between the forecast and observations is larger, which implies more categorical misses and/or false alarms at any SPI intensity. Notwithstanding, since the increase of SPI intensity is accompanied by a decrease of the respective cumulative probability, it was expected that the BSS would decrease with an increase of the SPI drought category because there is a larger probability for mismatching.

To finalize the evaluation of seasonal drought forecasts with the ECMWF S4 data set for South-Central America, we proceed with the analysis of the Relative Operating Characteristic (ROC) of the forecasts. In Figures 8 and 9, we present the spatial distribution of the area under the ROC curve for the probability of drought detection at different SPI frequencies. The values are estimated considering the ECMWF S4 SPI3 and SPI6 forecasts at a lead time of 3 and 6 months respectively, for the hindcast period. We have pooled together all seasons at each 1dd grid point in generating the maps of Figures 8 and 9. For the SPI3 and SPI6, for the "moderate" drought threshold, the area under the ROC curve at all grid points in South-Central America is well above the no skill line, indicating that, despite the poor reliability measured by the BSS, the forecasting system does have some skill. Nevertheless, similarly to the BSS, we perceive that the regions in the North of South America, Northeast of Argentina and Mexico are more skillful than the remaining locations. As the intensity of drought increases, the usefulness of the forecasting system decreases both in magnitude and area. For "extreme" drought events, the grid-points located in South, Central and Northeast of South America are not skillful, as the area under ROC curve is below the 0.5.

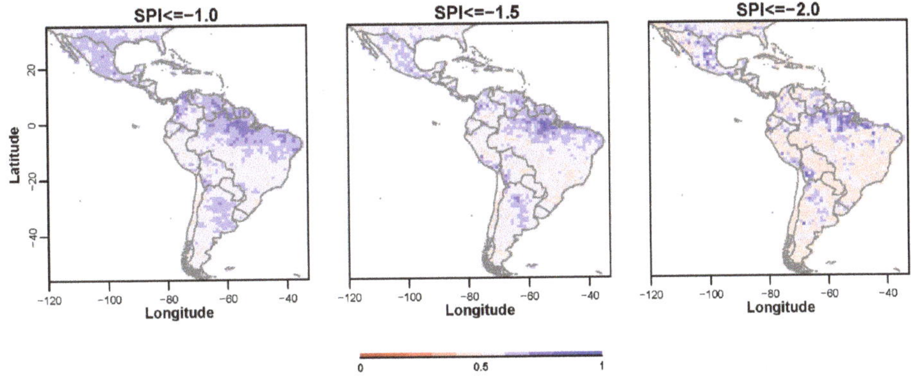

Figure 8. Area under the Relative Operating Characteristic (ROC) curve for the probability of drought detection at different SPI3 frequencies. Values indicated in the color bar are estimated at lead time of 3 months for the hindcast period 1981–2010.

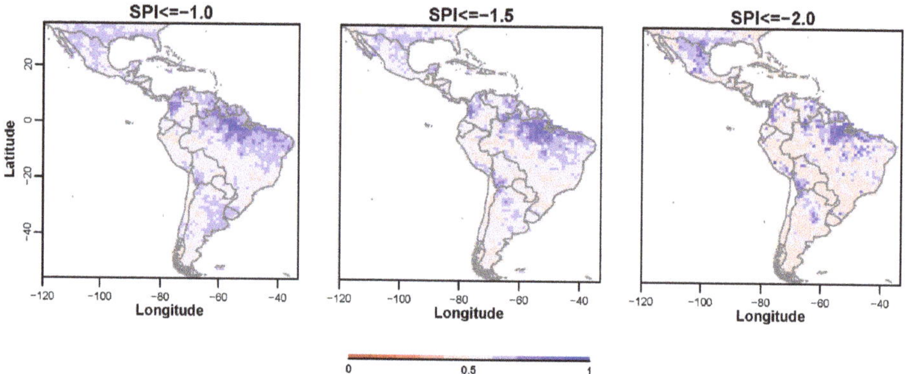

Figure 9. Area under the ROC curve for the probability of drought detection at different SPI6 frequencies. Values indicated in the color bar are estimated at lead time of 6 months for the hindcast period 1981–2010.

4. Conclusions

Here we present and assessment of seasonal drought forecasts, as characterized by the SPI at 3- and 6-months accumulation periods for, 3- and 6-month lead times, respectively. The main advantage of using the SPI for drought monitoring and prediction is that it is already used in operational monitoring systems in many countries around the globe and it is the drought index endorsed by the World Meteorological Organization (WMO).

We evaluated the scalar accuracy of the SPI forecasts together with the skill of probabilistic forecasts of discrete drought events (i.e., <-1). The skill of probabilistic drought identification with the SPI was also assessed. The scalar skill of the SPI-3 and SPI-6 was found to be seasonally and regionally dependent, but for some locations, SPI3 predictions at a lead of 3 months and SPI6 predictions at a lead of 6 months are found to have "useful" skill (monthly correlation with observations is statistically significant at the 5% significance level). The difference in skill between the ECMWF S4 SPI forecasts for South-Central America and a baseline forecast based on the climatological characteristics is positive in many areas and for many months, however it is mostly statistically insignificant. Nevertheless, for the SPI-3, our results show that the skill of the dynamic seasonal forecast is always equal to or above the climatological forecasts. On the other hand, for the SPI-6, our results indicate that it is more difficult to improve the climatological forecasts.

In a second step, we have evaluated several methods to forecast the drought events from the ensemble. Ensemble drought detection was based on several methods (Table A2) and can be organized into three types [40]: individual, where the index is based on an individual member or percentile; partially integrative, where the sum of particular individual members or percentiles are used; and integrative, which is represented by the ensemble mean. Although individual dry members and partially integrative methods were providing an outstanding accuracy for seasonal drought detection, our results have shown that the spread of the ensemble is too large and these methods also have large bias and false alarm ratio. The best (or most consistent) method is defined by using the ensemble mean SPI values, both for SPI3 and SPI6, at three and six months lead times. Our decision was based on the GSS index, which according to many authors provides an optimum solution for selection a classification method based on the number of hits, misses and false alarm ratio. The ensemble mean achieves an overall accuracy of about 80%, with POD above 30% for at least 75% of the study area, and false alarm ration that is overall below the 70%. Although the ECMWF S4 forecast system often overestimates the drought onset, it is significantly better than using the climatology ($\cong$16%).

Finally, standard verification measures for probabilistic forecasts were used to assess the accuracy of drought predictions based on the SPI values for "moderate", "severe" and "extreme" categories. The Brier Skill Score, which measures the probabilistic forecast skill against a forecast derived from the climatology, showed that both the SPI3 and SPI6 were, for some regions, slightly more skillful that the climatology. The ECMWF forecast system behaves better than the climatology for clustered grid points at the North of South America, Northeast of Argentina and Mexico. The skillful regions are similar for SPI-3 and -6, but become reduced in extent for the most severe SPI categories. We hypothesize that, because an increase of SPI intensity is accompanied by a decrease of the respective cumulative probability, the likelihood of mismatching is larger. As expected, the BSS is lower for the locations where the scalar mismatch between the forecast and the observations is larger, which implies more categorical misses and/or false alarms at any SPI intensity.

Forecasting different magnitudes of meteorological drought intensity on a seasonal time scale still remains a challenge. However, the ECMWF S4 forecasting system does capture reasonably well the onset of drought events (i.e., "moderate" drought) for some regions and seasons. A match is noticeable between observed and predicted SPI for dry months in arid regions with highly marked precipitation seasonality. Although the performance of Numerical Weather Prediction models is always improving and advances in the representation of physical processes in the models is an area of intense active research, the performance is still not good enough to provide useful guidance on months with high precipitation amounts; but it provides information that is more skillful than the climatology for dry periods.

Author Contributions: Conceptualization, H.C., G.N. and P.B.; Methodology, H.C., G.N. and C.L.; Data computation, H.C. and E.D., Validation, all authors; Writing-Original and Draft Preparation, All authors; Writing-Review & Editing, all authors.

Funding: This research received support from the EUROCLIMA regional cooperation program between the European Union (European Commission; DG DEVCO) and Latin America.

Conflicts of Interest: The authors declare no conflict of interest.

Appendix A Description of the Validation Metrics

Appendix A.1 Nonprobabilistic Forecasts of Continuous SPI Values

We first verify the scalar accuracy of the SPI values for the multimodel ensemble mean at 3 and 6 months lead time (respectively for SPI3 and SPI6). Ensemble mean SPI values are verified against observations for the hindcast period (i.e., from 1981 to 2010). In this case, the SPI magnitude can take any value in a specified segment of the real line, rather than being limited to a finite number of discrete classes (see Table A1). We perform an independent verification of drought forecasts for each month, by using four common accuracy measures of continuous nonprobabilistic forecasts, namely: the Pearson product-moment correlation coefficient, r; the Mean Error (ME); and the Root Mean Squared Error (RMSE). To be considered statistically significant at the 5% (10%) confidence level, the r between forecast SPI values and those in the verifying GPCC data needs to be greater than 0.37 (0.31), as defined by [41] for N_{years} = 29 observations (i.e., after subtracting 1 year from the total number of available years in the dataset).

Although the correlation does reflect linear association between two variables (in this case, forecasts and observations), it is not sensitive to biases that may be present in the forecasts. On the other hand, the ME, which is the difference between the average forecast and average observation, expresses the bias of the forecasts. Forecasts that are, on average, too high will exhibit ME > 0 and forecasts that are, on average, too low will exhibit ME < 0. It is important to note that the bias gives no information about the typical magnitude of individual forecast errors, and is therefore not in itself an accuracy measure. To complement the ME, we have computed the RMSE, which has the same physical dimensions as the forecasts and observations, and can be thought of as a typical mean magnitude for individual forecast errors.

We also verify the skill score for the multimodel ensemble mean at 3 and 6 months lead time (respectively for SPI3 and SPI6). Skill score refers to the relative accuracy of an ensemble set of forecasts and is interpreted as the improvement over a reference forecast [48]. Therefore, if the ECMWF S4 is providing value-added skill to the SPI forecasts, it will first be manifested by temporal correlations with observations, r_1, that exceed the expected correlation of the same observations with the climatological SPI baseline value (0), r_2. Under the assumption that the sets of forecasts are normally distributed, to assess the statistical significance of the difference between two correlations r_1 and r_2, we used Fisher's Z transformation, as explained in [49]. We define Z_i as

$$Z_i = \frac{1}{2}\ln\left(\frac{1+r_i}{1-r_i}\right)$$

for $i = 1$ and 2. The transformation Z is assumed to be normally distributed with variance $(N-3)-1$, where $N = 29$ observations (i.e., after subtracting 1 year from the total number of available years in the dataset). We then transformed r_1 and r_2 to Z_1 and Z_2, and computed the statistical significance for the difference in correlations using the Z statistics:

$$Z = \frac{Z_1 - Z_2}{\sqrt{\frac{1}{N_1-3} + \frac{1}{N_2-3}}}$$

where $N_1 - 3$ and $N_2 - 3$ are the degrees of freedom for r_1 and r_2, respectively. Using a null hypothesis of equal correlation and a non-directional alternative hypothesis of unequal correlation, if Z is greater than 1.96, the difference in correlations is statistically significant at the 5% confidence level.

A complementary skill score measure was constructed using the *RMSE* as the underlying accuracy statistic. The reference RMSE is based on the climatological average *SPI*, and is computed as:

$$RMSE_{Clim} = \sqrt{\frac{1}{N}\sum_{k=1}^{N}\left(\overline{SPI} - SPI_k\right)^2}$$

For the *SPI*, the climatological average does not change from forecast occasion to forecast occasion (i.e., as a function of the yearly index *k*). This implies that the $RMSE_{Clim}$ is an estimate of the sample variance of the predictand. For the RMSE using climatology as the control forecasts, the skill score becomes

$$SS_{Clim} = 1 - \frac{RMSE}{RMSE_{Clim}}$$

Because of the arrangement of the skill score, the SS_{Clim} based on RMSE is sometimes called the reduction of variance (RV), because the quotient being subtracted is the average squared error (or residual, in the nomenclature of regression) divided by the climatological variance.

Appendix A.2 Nonprobabilistic Forecasts of Categorical SPI Values

The temporal correlation between forecast and observed values of the SPI provides an overall measure of forecast accuracy and skill, one that is not limited to the case of drought alone. Therefore, we also evaluated SPI forecasts in the context of being able to detect drought, that is, when the SPI drops below a particular threshold. Here, we identified a drought event as occurring when the SPI value for a given month was ≤ -1, which corresponds to a "moderate drought" category or higher in the classification system presented in Table A1.

Table A1. SPI classification following McKee et al. [8].

SPI Value	Class	Cumulative Probability	Probability of Event [%]
SPI > 2.00	Extreme wet	0.977–1.000	2.3%
1.50 < SPI < 2.00	Severe wet	0.933–0.977	4.4%
1.00 < SPI < 1.50	Moderate wet	0.841–0.933	9.2%
−1.00 < SPI < 1.00	Near normal	0.159–0.841	68.2%
−1.50 < SPI < −1.00	Moderate dry	0.067–0.159	9.2%
−2.00 < SPI < −1.50	Severe dry	0.023–0.067	4.4%
SPI < −2.00	Extreme dry	0.000–0.023	2.3%

Ensemble drought detection was based on several methods (Table A2) and can be categorized into three types [50]: individual, where the index is based on an individual member or percentile; partially integrative, where the sum of particular individual members or percentiles are used; and integrative which is represented by the ensemble mean. The individual types should be seen as providing complementary information about the intensity of the SPI, but also about the distribution of the members. The individual types of drought detection have been subdivided into five classes representing dry members (Q13, Q23), wet ones (Q77, Q88) or the median. The extreme members of the distribution are not used to avoid outliers generally associated with ensemble systems [50].

Table A2. Methods to detect drought events from the S4 ensemble system. Adapted from [40].

Name	Definition	Type
13th percentile (Q13)	Member located at the 13% of the CDF	Individual
23th percentile (Q23)	Member located at the 23% of the CDF	Individual
Median (MED)	Member located at the 50% of the CDF	Individual
77th percentile (Q77)	Member located at the 77% of the CDF	Individual
88th percentile (Q88)	Member located at the 88% of the CDF	Individual
Large spread (SpL)	Sum of the extreme members (Q13 + Q88)	Partially integrative
Low spread (Spl)	Sum of the members (Q23 + Q78)	Partially integrative
Dry spread (SpD)	Sum of the dry members (Q13 + Q23)	Partially integrative
Flood spread (SpF)	Sum of the wet members (Q77 + Q88)	Partially integrative
Mean (EM_RES)	Ensemble mean	Integrative

For 3- and 6-month lead times (respectively for SPI3 and SPI6), we computed several verification measures for the categorical forecasts (i.e., below the SPI "-1" threshold) identified with the methods described in Table A2. All verification measures are based on a contingency table approach, which is applied at each grid point in the study area. The entries in the table are defined as follows: "A" is the number of drought events that are forecast and occur; "B" is the number of drought events that are forecast but do not occur; "C" is the number of drought events that are not forecast but do occur; and "D" is the number of drought events that are not forecast and do not occur. The variable N is the total number of cases analyzed from 1981 to 2010. Based on these values, the percentage correct (PC, perfect = 1) is the ratio of good forecasting events in relation to the total number of events.

$$PC = \frac{A + D}{N}$$

The extreme dependency score (EDS) provides a skill score in the range [−1, 1] that can be used to find the hit-rate exponent [51]. The EDS takes the value of 1 for perfect forecasts and 0 for random forecasts, and is greater than zero for forecasts that have hit rates that converge slower than those of random forecasts.

$$EDS = \frac{2log\frac{A+B}{N}}{log\frac{A}{N}} - 1$$

The Gilbert skill score (GSS) measures the fraction of forecast events that were correctly predicted, adjusted for the frequency of hits that would be expected to occur simply by random chance [40].

$$GSS = \frac{A + A^*}{A + B + C - A^*}$$

where A^* is the number of random hits, computed as:

$$A^* = \frac{(A + B)(A + C)}{N}$$

The GSS is often used in the verification of rainfall forecasts because its "equitability" allows scores to be compared more fairly across different regimes (for example, it is easier to correctly forecast rain occurrence in a wet climate than in a dry climate). However, because it penalizes both misses and false alarms in the same way, it does not distinguish the source of forecast error. Therefore, it should be used in combination with at least one other contingency table statistic, for example, bias. Here, we compute bias as:

$$BIAS = \frac{A + B}{A + C}$$

The probability of detection (POD, perfect = 1) is the ratio of the total number of observed events that have been forecasted.

$$POD = \frac{A}{A + C}$$

The false alarm rate (FAR, perfect = 0) is the fraction of the forecasted events which actually did not occur.

$$FAR = \frac{B}{A + B}$$

Appendix A.3 Probabilistic Forecast of Categorical SPI Values

Verification of probability forecasts is somewhat more subtle than verification of non-probabilistic forecasts. Since non-probabilistic forecasts contain no expression of uncertainty, it is clear whether an individual forecast is correct or not. On the other hand, unless a probabilistic forecast is either 0.0 or 1.0, the situation is less clear-cut. For probability values between these two (certainty) extremes, a forecast is neither right nor wrong, so that meaningful assessments can only be made using collections of multiple forecast members and observation pairs. A number of accuracy measures for verification of probabilistic forecasts of dichotomous events exist, but by far the most common is the Brier score (BS) [48]. The Brier score is essentially the mean squared error of the probability forecasts, considering that the GPCC drought observation at time k is $o_k = 1$ if a drought event occurs (i.e., SPI ≤ -1), and that the GPCC observation at time k is $o_k = 0$ if a drought event does not occur (i.e., SPI > -1). The BS averages the squared differences between pairs of forecast probabilities, $fcst_k$, and the subsequent binary reference observations,

$$BS = \frac{1}{n}\sum_{k=1}^{N}(fcst_k - o_k)^2$$

where the index k again denotes a numbering of the N forecast-event pairs. Comparing the BS with the root-mean squared error, it can be seen that the two are completely analogous. As a mean-squared-error measure of accuracy, the BS is negatively oriented, with perfect forecasts exhibiting $BS = 0$. Less accurate forecasts receive higher BS values, but since individual forecasts and observations are both bounded by zero and one, the score can take on values only in the range $0 \leq BS \leq 1$.

Skill scores of the form of SS_{clim} are also often computed for the BS, yielding the Brier Skill Score (BSS):

$$BSS = 1 - \frac{BS}{BS_{ref}}$$

The BSS is the conventional skill-score form using the BS as the underlying accuracy measure. Usually, for the SPI, the reference forecasts are the relevant climatological probabilities of a drought event taking place with a certain severity (Table A1). For example, the frequency of "Moderate" drought events is approximately the 16%. The BSS ranges between minus infinity and 1; 0 indicates no skill when compared to the reference forecast; the perfect score is 1. A good companion to the BSS is the Relative Operating Characteristic (ROC) of the forecast. ROC is conditioned on the observations, and measures the ability of the probabilistic forecasting system to discriminate between drought events and non-events of different frequencies, that is, the resolution of the forecast. ROC is not sensitive to bias in the forecast (even a biased forecast could give a good ROC). However, the ROC is a measure of potential usefulness of the probabilistic forecast, and the area under the ROC curve gives a measure of its skill. Since ROC curves for perfect forecasts pass through the upper-left corner, the area under a perfect ROC curve includes the entire unit square, so $A_{perf} = 1$. Similarly ROC curves for random forecasts lie along the 45° diagonal of the unit square, yielding the area $A_{rand} = 0.5$. The area A under a ROC curve of interest can also be expressed in standard skill-score form SS_{ROC}, as

$$SS_{ROC} = \frac{A - 1/2}{1 - 1/2}$$

Wilks [48] states that SS_{ROC} is a reasonably good discriminator among relatively low-quality forecasts, but that relatively good forecasts tend to be characterized by quite similar (near-unit) areas under their ROC curves. The SS_{ROC} ranges between 0 and 1; 0.5 indicates no skill, while the perfect score is 1.

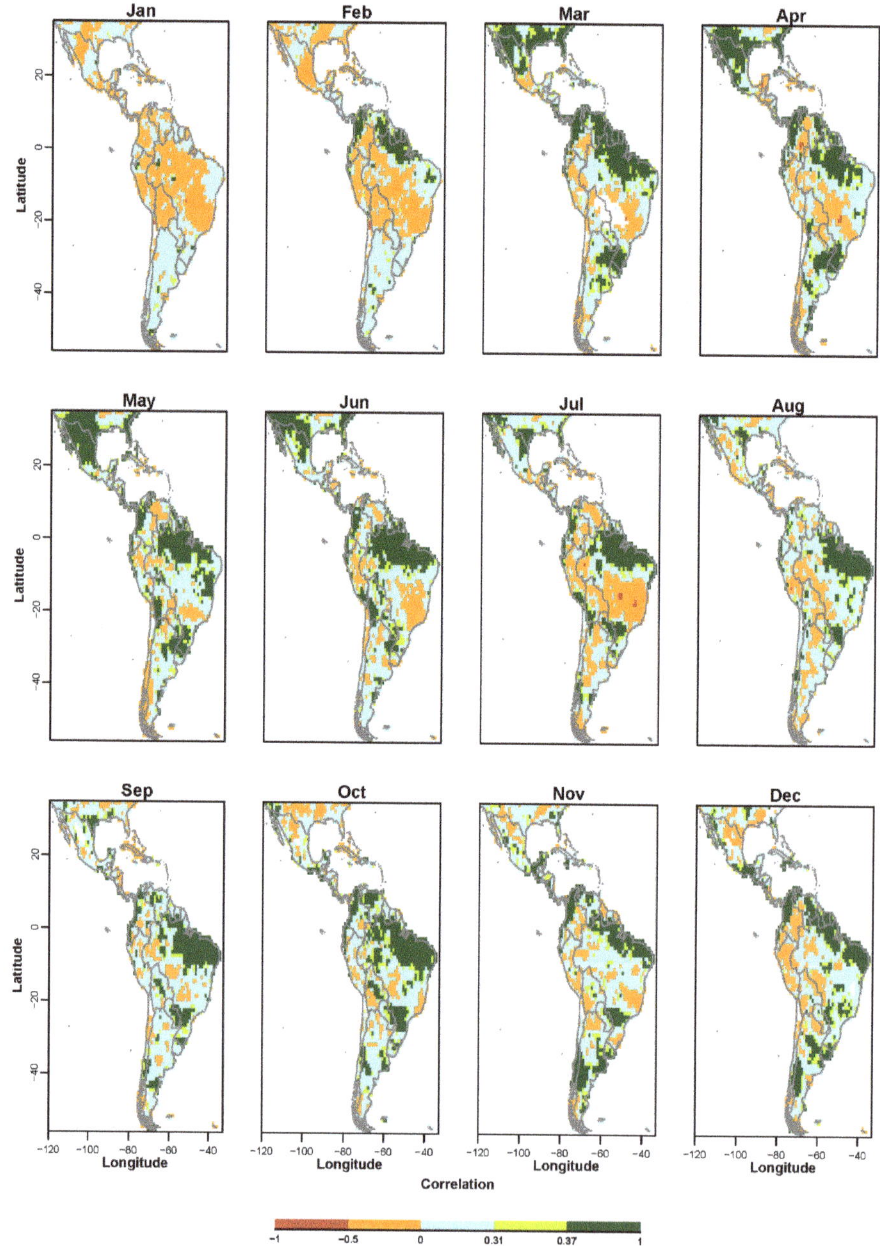

Figure A1. Monthly correlation of the observed and forecast SPI at 6-months lead time (SPI6) (using the mean of the ensemble) for the hindcast period (1981–2010). Values are indicated in the color bar: 0.31 (0.37) is statistical significant at 10% (5%) significance level.

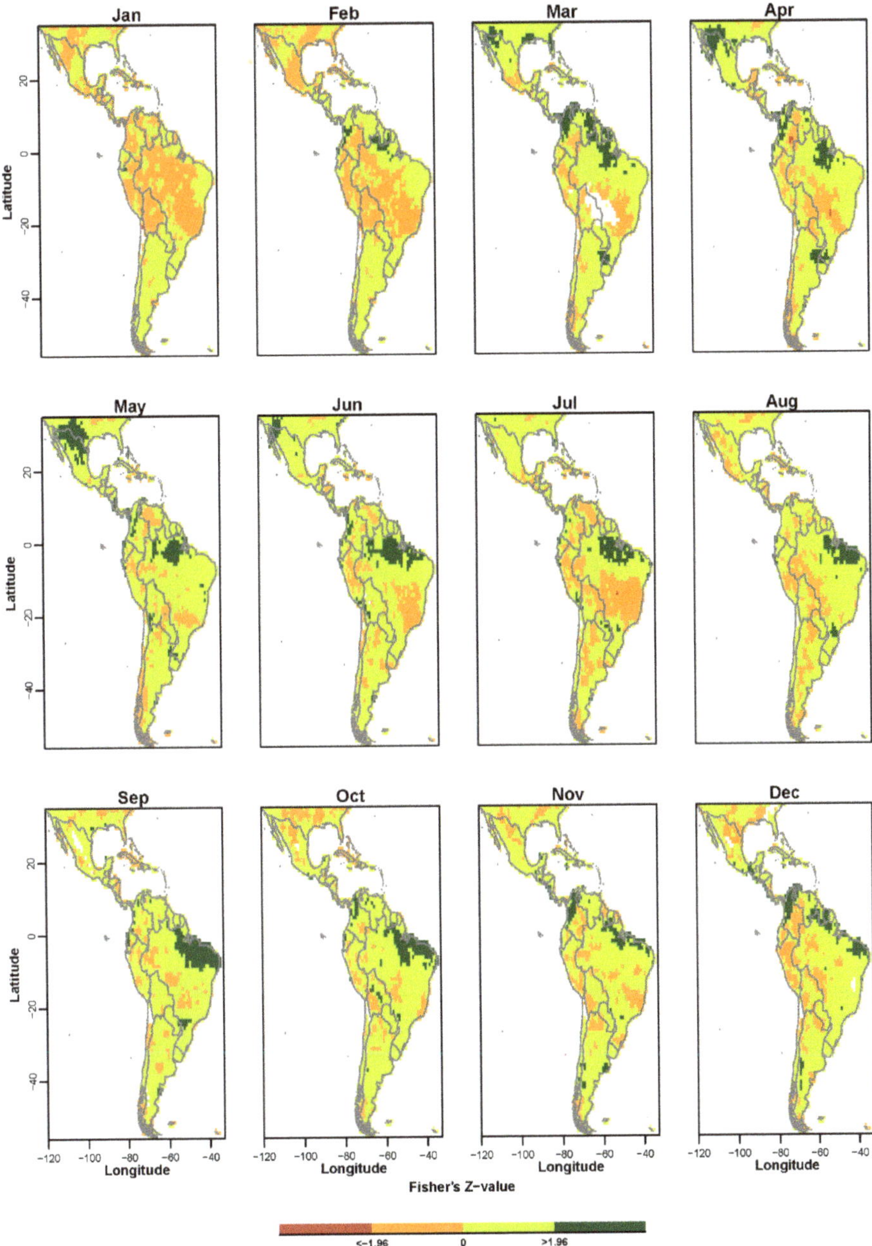

Figure A2. Monthly difference in forecast skill (Pearson correlation) between the forecast SPI6 at 6-month lead time (using the mean of the ensemble) and climatological SPI for the hindcast period (1981–2010). Values are indicated in the color bar: 1.96 is the statistical significant at the 5% significance level.

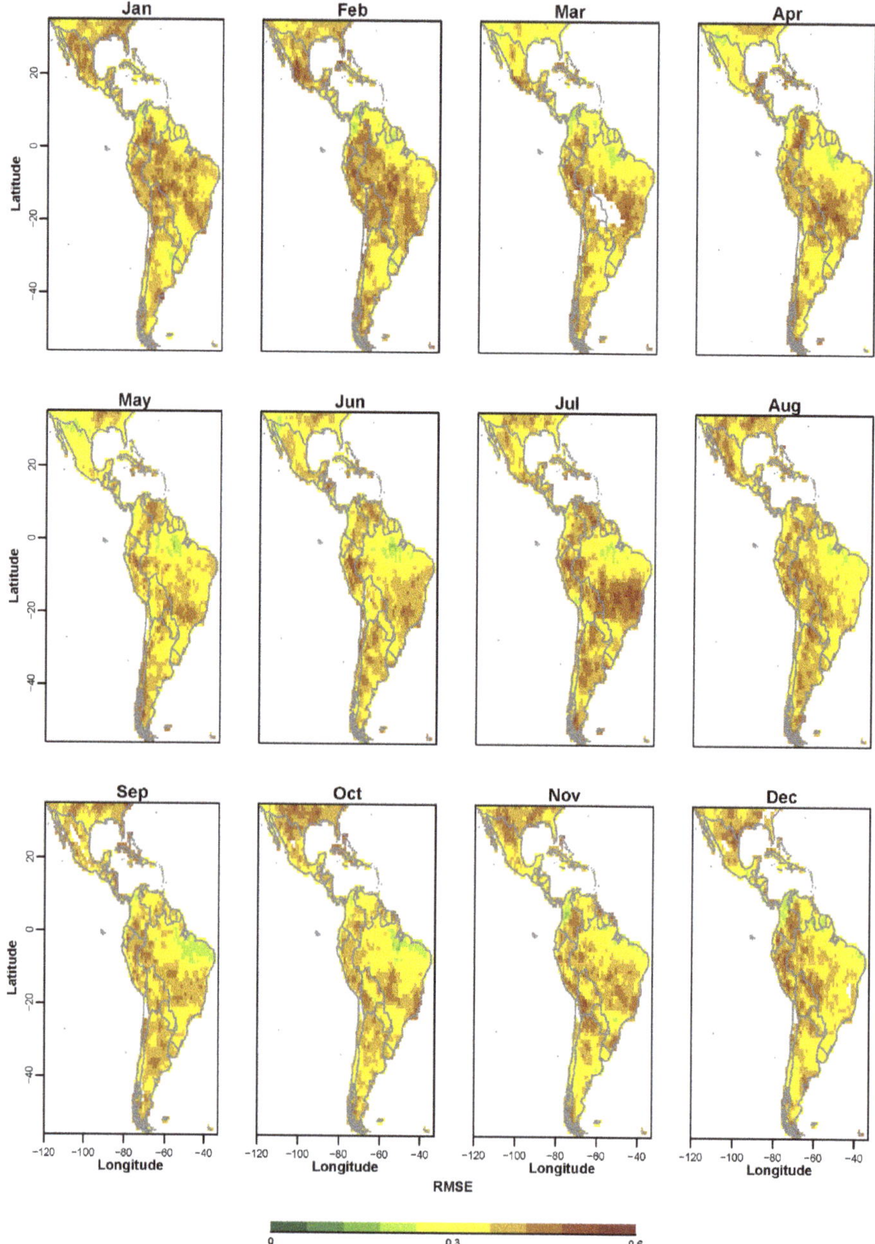

Figure A3. RMSE between the observed and forecast SPI6 at 6-month lead time (mean of the ensemble) for the hindcast period (1981–2010). Values in difference of percentile magnitude are indicated in the color bar.

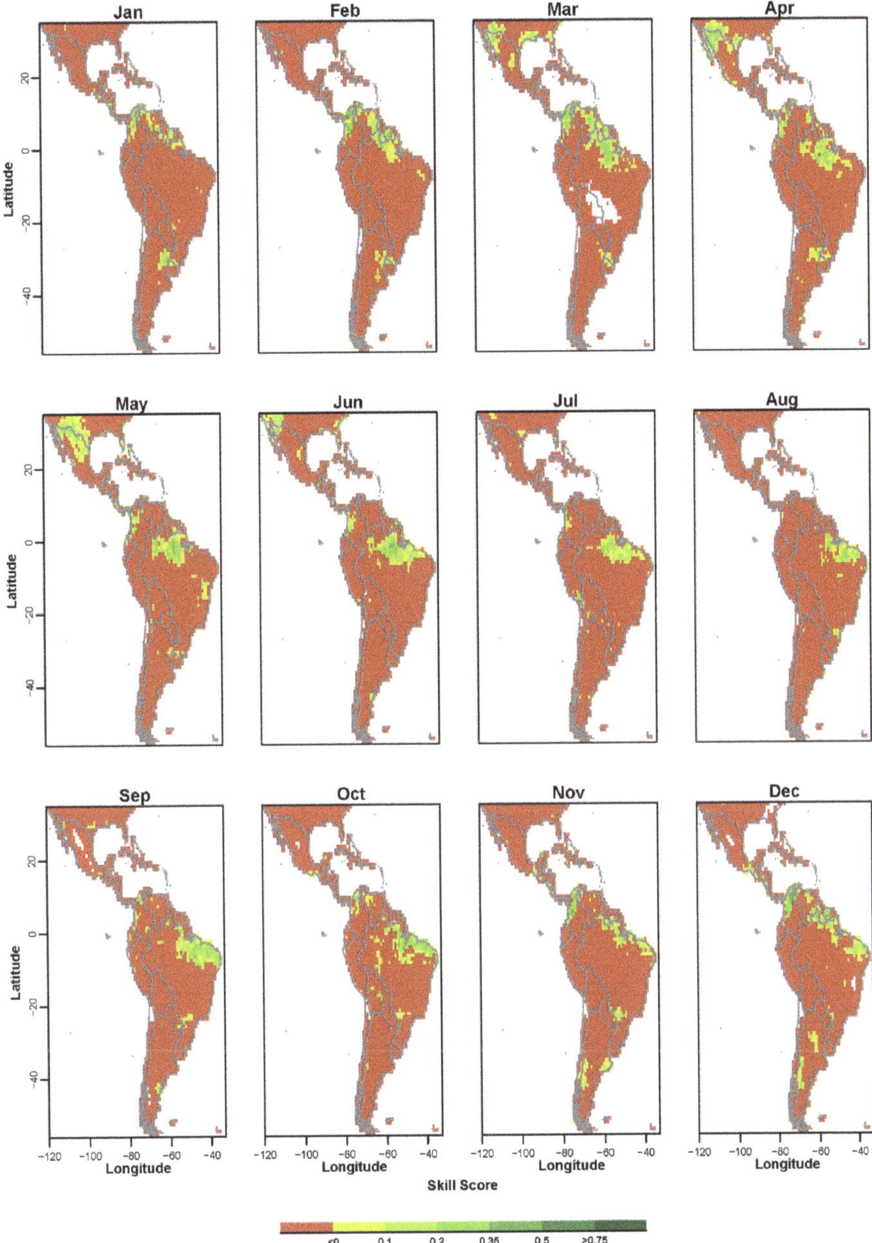

Figure A4. Skill Score of the SPI6 at 6-month lead time forecast measured in terms of the RMSE relative to climatological RMSE for the hindcast period (1981–2010).

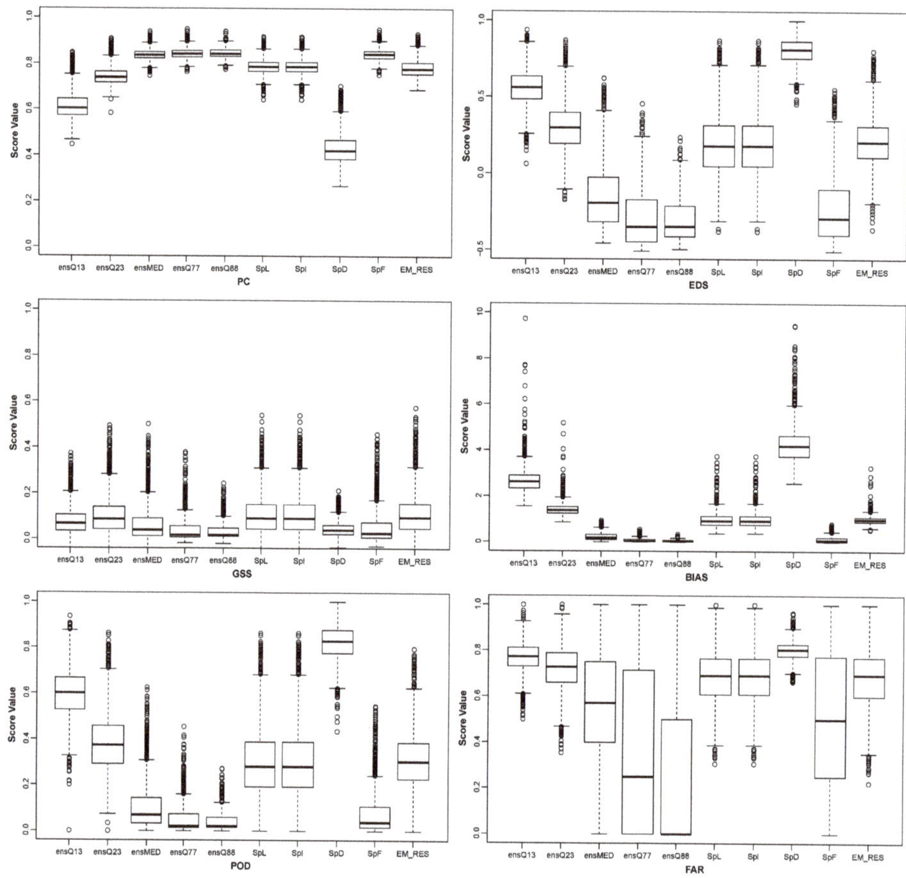

Figure A5. Verification measures of categorical drought forecasts (i.e., below the SPI6 "-1" threshold) estimated with the methods described in Table A2.

References

1. Carrão, H.; Singleton, A.; Naumann, G.; Barbosa, P.; Vogt, J. An optimized system for the classification of meteorological drought intensity with applications in frequency analysis. *J. Appl. Meteorol. Climatol.* **2014**, *53*, 1943–1960. [CrossRef]
2. Goddard, S.; Harms, S.K.; Reichenbach, S.E.; Tadesse, T.; Waltman, W.J. Geospatial decision support for drought risk management. *Commun. ACM* **2003**, *46*, 35–37. [CrossRef]
3. Dai, A. Drought under global Warming: A review. *Wiley Interdiscip. Rev. Clim. Chang.* **2011**, *2*, 45–65. [CrossRef]
4. Lloyd-Hughes, B. The impracticality of a universal drought definition. *Theor. Appl. Climatol.* **2014**, *117*, 607–611. [CrossRef]
5. Steinemann, A.C.; Cavalcanti, L.F. Developing multiple indicators and triggers for drought plans. *J. Water Res. Plan. Manag.* **2006**, *132*, 164–174. [CrossRef]
6. Heim, R.R. A review of twentieth-century drought indices used in the United States. *Bull. Am. Meteorol. Soc.* **2002**, *83*, 1149–1165. [CrossRef]
7. Vicente-Serrano, S.M.; Beguerıa, S.; Lopez-Moreno, J.I. A multiscalar drought index sensitive to global warming: The standardized precipitation evapotranspiration index. *J. Clim.* **2010**, *23*, 1696–1718. [CrossRef]

8. McKee, T.B.; Doeskin, N.J.; Kleist, J. The relationship of drought frequency and duration to time scales. In Proceedings of the 8th Conference on Applied Climatology, American Meteorological Society, Anaheim, CA, USA, 17–22 January 1993; pp. 179–184.
9. Svoboda, M.; Hayes, M.; Wood, D. *Standardized Precipitation Index User Guide*; WMO-No. 1090; World Meteorological Organization (WMO): Geneva, Switzerland, 2012.
10. Mishra, A.K.; Singh, V.P. A review of drought concepts. *J. Hydrol.* **2010**, *391*, 202–216. [CrossRef]
11. Kim, T.-W.; Valds, J.B. Nonlinear model for drought forecasting based on a conjunction of wavelet transforms and neural networks. *J. Hydrol. Eng.* **2003**, *8*, 319–328. [CrossRef]
12. Mishra, A.K.; Desai, V.R.; Singh, V.P. Drought forecasting using a hybrid stochastic and neural network model. *J. Hydrol. Eng.* **2007**, *12*, 626–638. [CrossRef]
13. Mishra, A.K.; Desai, V.R. Drought forecasting using stochastic models. *Stoch. Environ. Res. Risk Assess.* **2005**, *19*, 326–339. [CrossRef]
14. Vitart, F.; Buizza, R.; Alonso Balmaseda, M.; Balsamo, G.; Bidlot, J.-R.; Bonet, A.; Fuentes, M.; Hofstadler, A.; Molteni, F.; Palmer, T.N. The new VAREPS-monthly forecasting system: A first step towards seamless prediction. *Q. J. R. Meteorol. Soc.* **2008**, *134*, 1789–1799. [CrossRef]
15. Nijssen, B.; Shukla, S.; Lin, C.; Gao, H.; Zhou, T.; Ishottama; Sheffield, J.; Wood, E.F.; Lettenmaier, D.P. A prototype global drought information system based on multiple land surface models. *J. Hydrometeorol.* **2014**, *15*, 1661–1676. [CrossRef]
16. Yuan, X.; Wood, E.F. Multimodel seasonal forecasting of global drought onset. *Geophys. Res. Lett.* **2013**, *40*, 4900–4905. [CrossRef]
17. Hao, Z.; AghaKouchak, A.; Nakhjiri, N.; Farahmand, A. Global integrated drought monitoring and prediction system. *Sci. Data* **2014**, *1*, 140001. [CrossRef] [PubMed]
18. Dutra, E.; Wetterhall, F.; Di Giuseppe, F.; Naumann, G.; Barbosa, P.; Vogt, J.; Pozzi, W.; Pappenberger, F. Global meteorological drought Part 1: Probabilistic monitoring. *Hydrol. Earth Syst. Sci.* **2014**, *18*, 2657–2667. [CrossRef]
19. Dutra, E.; Pozzi, W.; Wetterhall, F.; Di Giuseppe, F.; Magnusson, L.; Naumann, G.; Barbosa, P.; Vogt, J.; Pappenberger, F. Global meteorological drought Part 2: Seasonal forecasts. *Hydrol. Earth Syst. Sci.* **2014**, *18*, 2669–2678. [CrossRef]
20. Spennemann, P.C.; Rivera, J.A.; Osman, M.; Saulo, A.C.; Penalba, O.C. Assessment of seasonal soil moisture forecasts over Southern South America with emphasis on dry and wet events. *J. Hydrometeorol.* **2017**, *18*, 2297–2311. [CrossRef]
21. Sheffield, J.; Andreadis, K.M.; Wood, E.F.; Lettenmaier, D.P. Global and continental drought in the second half of the twentieth century: Severity–area–duration analysis and temporal variability of large-scale events. *J. Clim.* **2009**, *22*, 1962–1981. [CrossRef]
22. Zhao, M.; Running, S.W. Drought-induced reduction in global terrestrial net primary production from 2000 through 2009. *Science* **2010**, *329*, 940–943. [CrossRef] [PubMed]
23. Vargas, W.M.; Naumann, G.; Minetti, J.L. Dry spells in the river Plata Basin: An approximation of the diagnosis of droughts using daily data. *Theor. Appl. Climatol.* **2011**, *104*, 159–173. [CrossRef]
24. Field, C.B.; Barros, V.; Stocker, T.F.; Qin, D.; Dokken, D.J.; Ebi, K.L.; Mastrandrea, M.D.; Mach, K.J.; Plattner, G.-K.; Allen, S.K.; et al. (Eds.) *Managing the Risks of Extreme Events and Disasters to Advance Climate Change Adaptation*; Cambridge University Press: New York, NY, USA, 2012; pp. 1–19.
25. Penalba, O.C.; Rivera, J.A. Future changes in drought characteristics over Southern South America projected by a CMIP5 multi-model ensemble. *Am. J. Clim. Chang.* **2013**, *2*, 173–182. [CrossRef]
26. Trenberth, K.E.; Stepaniak, D. Indices of El Niño evolution. *J. Clim.* **2011**, *14*, 1697–1701. [CrossRef]
27. FAO. Aquastat Database. Food and Agriculture Organization of the United Nations (FAO). 2017. Available online: http://www.fao.org/nr/water/aquastat/main/index.stm (accessed on 15 October 2017).
28. Magrin, G.; Garcia, C.G.; Choque, D.C.; Gimenez, J.C.; Moreno, A.R.; Nagy, G.J.; Nobre, C.; Villamizar, A. Climate change, Impacts, adaptation and vulnerability. In *Contribution of Working Group II to the Fourth Assessment Report of the Intergovernmental Panel on Climate Change*; Parry, M.L., Canziani, O.F., Palutikof, J.P., van der Linden, P., Hanson, C.E., Eds.; Cambridge University Press: Cambridge, UK, 2007; pp. 581–615.
29. Llano, M.P.; Vargas, W.; Naumann, G. Climate variability in areas of the world with high production of soya beans and Corn: Its relationship to crop yields. *Meteorol. Appl.* **2012**, *19*, 385–396. [CrossRef]

30. Molteni, F.; Stockdale, T.; Balmaseda, M.; Balsamo, G.; Buizza, R.; Ferranti, L.; Magnunson, L.; Mogensen, K.; Palmer, T.; Vitart, F. The new ECMWF seasonal forecast system (System 4). In *Technical Memorandum No. 656*; European Centre for Medium Range Weather Forecasts: Berkshire, UK, 2011.
31. Rudolf, B.; Becker, A.; Schneider, U.; Meyer-Christoffer, A.; Ziese, M. New full data reanalysis version 5 provides high-quality gridded monthly precipitation data. *GEWEX News* **2011**, *21*, 4–5.
32. Naumann, G.; Barbosa, P.; Carrão, H.; Singleton, A.; Vogt, J. Monitoring drought conditions and their uncertainties in Africa using TRMM data. *J. Appl. Meteorol. Climatol.* **2012**, *51*, 1867–1874. [CrossRef]
33. Svoboda, M.; LeComte, D.; Hayes, M.; Heim, R.; Gleason, K.; Angel, J.; Rippey, B.; Tinker, R.; Palecki, M.; Stooksbury, D.; et al. The drought monitor. *Bull. Am. Meteorol. Soc.* **2002**, *83*, 1181–1190. [CrossRef]
34. Steinemann, A. Drought indicators and Triggers: A stochastic approach to evaluation. *JAWRA* **2003**, *39*, 1217–1233. [CrossRef]
35. Hayes, M.J.; Svoboda, M.D.; Wilhite, D.A.; Vanyarkho, O.V. Monitoring the 1996 drought using the standardized precipitation index. *Bull. Am. Meteorol. Soc.* **1999**, *80*, 429–438. [CrossRef]
36. Wu, H.; Hayes, M.J.; Wilhite, D.A.; Svoboda, M.D. The effect of the length of record on the standardized precipitation index calculation. *Int. J. Climatol.* **2005**, *25*, 505–520. [CrossRef]
37. Sepulcre-Canto, G.; Horion, S.; Singleton, A.; Carrão, H.; Vogt, J. Development of a combined drought indicator to detect agricultural drought in Europe. *Earth Syst. Sci.* **2012**, *12*, 3519–3531. [CrossRef]
38. Ntale, H.K.; Gan, T.Y. Drought indices and their application to east Africa. *Int. J. Climatol.* **2003**, *23*, 1335–1357. [CrossRef]
39. Hofer, B.; Carrao, H.; Mcinerney, D. Multi-disciplinary forest fire danger assessment in Europe: The potential to integrate long-term drought information. *IJSDIR* **2012**, *7*, 300–322.
40. Lavaysse, C.; Vogt, J.; Pappenberger, F. Early warning of drought in Europe using the monthly ensemble system from ECMWF. *Hydrol. Earth Syst. Sci.* **2015**, *19*, 3273–3286. [CrossRef]
41. Mo, K.C.; Lyon, B. Global meteorological drought prediction using the North American multi-model ensemble. *J. Hydrometeorol.* **2015**, *16*, 1409–1424. [CrossRef]
42. Vera, C.S.; Alvarez, M.S.; Gonzalez, P.L.; Liebmann, B.; Kiladis, G.N. Seasonal cycle of precipitation variability in South America on intraseasonal timescales. *Clim. Dyn.* **2017**, 1–11. [CrossRef]
43. González, P.L.; Vera, C.S. Summer precipitation variability over South America on long and short intraseasonal timescales. *Clim. Dyn.* **2014**, *43*, 1993–2007. [CrossRef]
44. González, P.L.; Vera, C.S.; Liebmann, B.; Kiladis, G. Intraseasonal variability in subtropical South America as depicted by precipitation data. *Clim. Dyn.* **2008**, *30*, 727–744. [CrossRef]
45. Ghelli, A.; Primo, C. On the use of the extreme dependency score to investigate the performance of an NWP model for rare events. *Meteorol. Appl.* **2009**, *16*, 537–544. [CrossRef]
46. Ruscica, R.C.; Sörensson, A.A.; Menéndez, C.G. Pathways between soil moisture and precipitation in southeastern South America. *Atmos. Sci. Lett.* **2015**, *16*, 267–272. [CrossRef]
47. Spennemann, P.C.; Saulo, A.C. An estimation of the land–atmosphere coupling strength in South America using the Global Land Data Assimilation System. *Int. J. Climatol.* **2015**, *35*, 4151–4166. [CrossRef]
48. Wilks, D.S. *Statistical Methods in the Atmospheric Sciences*, 2nd ed.; Academic Press: Amsterdam, The Netherlands, 2005.
49. Quan, X.W.; Hoerling, M.P.; Lyon, B.; Kumar, A.; Bell, M.A.; Tippett, M.K.; Wang, H. Prospects for dynamical prediction of meteorological drought. *J. Appl. Meteorol. Climatol.* **2012**, *51*, 1238–1252. [CrossRef]
50. Lavaysse, C.; Carrera, M.; Blair, S.; Gagnon, N.; Frenette, R.; Charron, M.; Yau, M.K. Impact of surface parameter uncertainties within the Canadian regional ensemble prediction system. *Mon. Weather Rev.* **2013**, *141*, 1506–1526. [CrossRef]
51. Stephenson, D.B.; Casati, B.; Ferro, C.A.T.; Wilson, C.A. The extreme dependency Score: A non-vanishing measure for forecasts of rare events. *Meteorol. Appl.* **2008**, *15*, 41–50. [CrossRef]

MDPI
St. Alban-Anlage 66
4052 Basel
Switzerland
Tel. +41 61 683 77 34
Fax +41 61 302 89 18
www.mdpi.com

Climate Editorial Office
E-mail: climate@mdpi.com
www.mdpi.com/journal/climate

www.ingramcontent.com/pod-product-compliance
Lightning Source LLC
LaVergne TN
LVHW070642170726
843515LV00004B/307